Practical Radio–Frequency Handbook

Practical Radio-Frequency Handbook
Second edition

IAN HICKMAN, BSc (Hons), CEng, MIEE, MIEEE

 Newnes

Newnes
An imprint of Butterworth-Heinemann
Linacre House, Jordan Hill, Oxford OX2 8DP
A division of Reed Educational and Professional Publishing Ltd

℞ A member of the Reed Elsevier plc group

OXFORD   BOSTON   JOHANNESBURG
MELBOURNE   NEW DELHI   SINGAPORE

First published 1993 as *Newnes Practical RF Handbook*
Second edition 1997

British Library Cataloguing in Publication Data
Hickman, Ian
  Practical Radio-Frequency Handbook
  I. Title
  621.384

ISBN 0 7506 34472

Library of Congress Cataloguing-in-Publication Data
Hickman, Ian.
  Practical Radio-Frequency Handbook/Ian Hickman.
  p.   cm.
  Includes bibliographical references and index.
  ISBN 0 7506 3447 2
  1. Radio circuits – Design and construction – Handbooks, manuals,
  etc.   I. Title.
  TK6554.H49 1993
  621.384′12–dc20   92–34846
                                                              CIP

Typeset by 🅐 Tek-Art, Croydon, Surrey
Printed and bound in Great Britain by
Biddles Ltd, Guildford and King's Lynn

Contents

Preface   vii

Acknowledgements   x

1   **Passive components and circuits**   1
    Resistance and resistors   1
    Capacitors   4
    Inductors and transformers   7
    Passive circuits   11

2   **RF transmission lines**   21

3   **RF transformers**   27

4   **Couplers, hybrids and directional couplers**   44

5   **Active components for RF uses**   52

6   **RF small-signal amplifiers**   73

7   **Modulation and demodulation**   85

8   **Oscillators**   103

9   **RF power amplifiers**   129
    Safety hazards to be considered   129
    First design decisions   130
    Levellers, VSWR protection, RF routing switches   131
    Starting the design   131
    Low-pass filter design   132
    Discrete PA stages   135

10   **Transmitters and receivers**   160

11   **Propagation**   177

12   **Antennas**   190

13   Attenuators and equalizers 212

14   Measurements 219
Measurements on CW signals 219
Modulation measurements 221
Spectrum and network analysers 223
Other instruments 227

Appendix 1   Useful relationships 234

Appendix 2   S-Parameters 241

Appendix 3   Attenuators (pads) 246

Appendix 4   Universal resonance curve 248

Appendix 5   RF cables 249

Appendix 6   Wire gauges and useful information 253

Appendix 7   Ferrite manufacturers 255

Appendix 8   Types of modulation — classification 257

Appendix 9   Quartz crystals 259

Appendix 10   Elliptic filters 261

Appendix 11   Screening 273

Appendix 12   Worldwide minimum external noise levels 279

Appendix 13   Frequency allocations 283

Appendix 14   SRDs (short range devices) 292

Index 294

Preface

The *Practical Radio–Frequency Handbook* aims to live up to its title, as a useful *vade-mecum* and companion for all who wish to extend their familiarity with RF technology. It is hoped that it will prove of use to practising electronic engineers who wish to move into the RF design area, or who have recently done so, and to engineers, technicians, amateur radio enthusiasts, electronics hobbyists and all with an interest in electronics applied to radio frequency communications. From this, you will see that it is not intended to be a textbook in any shape or form. Nothing would have been easier than to fill it up with lengthy derivations of formulae, but readers requiring to find these should look elsewhere. Where required, formulae will be found simply stated: they are there to be used, not derived.

I have naturally concentrated on current technology but have tried to add a little interest and colour by referring to earlier developments by way of background information, where this was thought appropriate, despite the pressure on space. This pressure has meant that, given the very wide scope of the book (it covers devices, circuits, equipment, systems, radio propagation and external noise), some topics have had to be covered rather more briefly than I had originally planned. However, to assist the reader requiring more information on any given topic, useful references for further reading are included at the end of most chapters. The inclusion of descriptions of earlier developments is by no means a waste of precious space for, in addition to adding interest, these earlier techniques have a way of reappearing from time to time — especially in the current climate of deregulation. A good example of this is the super-regenerative receiver, which appeared long before the Second World War, did sterling service during that conflict, but was subsequently buried as a has-been: it is now reappearing in highly price-sensitive short-range applications such as remote garage door openers and central locking controllers.

Good RF engineers are currently at a premium, and I suspect that they always will be. The reason is partly at least to be found in the scant coverage which the topic receives in university and college courses. It is simply so much easier to teach digital topics, which furthermore — due to the rapid advances being made in the technology — have long seemed the glamorous end of the business. However, the real world is analogue, and communicating information, either in analogue or digital form, at a distance and without wires, requires the use of electromagnetic radiation. This may be RF, microwave, millimetre wave or optical and there is a whole technology associated with each. This book deals just with the RF portion of the spectrum, which I have taken as extending up to 1000 MHz. This is used for an enormous variety of services, including sound broadcasting and television, commercial, professional, government and military communications of all kinds, telemetry and telecontrol, radio telex and facsimile and amateur radio. There are specialized applications, such as short-range communications and control (e.g. radio microphones, garage door openers) and a number of more sinister applications such as ESM, ECM and ECCM (electronic surveillance measures, e.g. eavesdropping; electronic counter measures, e.g. exploitation and jamming; and electronic counter counter measures, e.g. jamming resistant radios using frequency hopping or direct sequence spread spectrum). Indeed, the pressure on spectrum space has never been greater than it is now and it is people with a knowledge of RF who have to design, produce, maintain and use equipment capable of working in this crowded environment. It is hoped that this book will prove useful to those engaged in these tasks.

This second edition has numerous minor additions and some corrections throughout, but the major change concerns the original Chapter 11, Antennas and propagation. In this edition, this chapter has been split into two. The new Chapter 11 deals solely with propagation and incorporates a great deal of extra material. A new Chapter 12 deals exclusively with antennas, and again, the opportunity has been taken to include a great deal of extra information.

No less important is the addition of three new appendices. Apendix 12 gives information, drawn from an ITU-R Report, about external noise levels — and incidentally some information about the ITU itself. Appendix 13 gives details of frequency allocations. Annexe 1 covers the documents defining UK frequency allocations. These documents are too large to reproduce in their entirety within the covers of this volume, but sample pages are included to give an idea of the wealth of information they include. Complete copies and further information may be obtained from the address given in the appendix. Annexe 2 likewise gives brief details of frequency allocations in the USA. Appendix 14 gives information relating to low power, short range radio devices. These represent an explosive area of growth at the present time, for a number of reasons. First, many of these devices require no licence — a great convenience to the end user —

although naturally the manufacturer must ensure that such a device meets the applicable specification. Second, due to the very limited range, frequencies can be re-used almost without limit, in a way not possible in, for example, broadcast applications, or even in PMR (private mobile radio). Details of the relevant specifications are found in Appendix 14.

It is hoped that the additions and alterations incorporated in this second edition will make the work even more useful to all with an interest in RF technology.

Ian Hickman

Acknowledgements

My thanks are due to my colleagues C.W. (appropriate initials!) who was largely responsible for Chapter 9, and M.H.G. who vetted and helpfully suggested many improvements to Chapter 11.

My thanks are also due to all the following, for providing illustrations or for permission to reproduce material supplied by them.

Anritsu Europe Ltd
Electronics World and Wireless World
GEC Plessey Semiconductors Ltd
Hewlett Packard Co.
Institute of Electrical and Electronics Engineers
Institution of Electrical Engineers
Marconi Instruments Ltd
Motorola Inc.
Motorola European Cellular Subscriber Division
Racal Antennas Ltd
Racal Communications Ltd
RFI Shielding Ltd
SEI Ltd
Siemens Plessey Defence Systems Ltd
Skandinavisk Teleindustri SKANTI A/S
Transradio Ltd

1
Passive components and circuits

The passive components used in electronic circuits all make use of one or more of the three fundamental phenomena of resistance, capacitance and inductance. Some components depend for their operation on the interaction between one of these electrical properties and a mechanical property, e.g. crystals used as frequency standards, piezo-electric sounders, etc. The following sections look at components particularly in the light of their suitability for use at RFs, and at how they can be inter-connected for various purposes.

Resistance and resistors

Some substances conduct electricity well; these substances are called *conductors*. Others called *insulators*, such as glass, polystyrene, wax, PTFE, etc., do not, in practical terms, conduct electricity at all: their resistivity is about 10^{18} times that of metals. Even though metals conduct electricity well, they still offer some resistance to the passage of an electric current, which results in the dissipation of heat in the conductor. In the case of a wire of length l metres and cross-sectional area A square metres, the current I in amperes which flows when an electrical supply with an electromotive force (EMF) of E volts is connected across it is given by $I = E/((l/A)\rho)$, where ρ is a property of the material of the wire, called *resistivity*. The term $(l/A)\rho$ is called the resistance of the wire, denoted by R, so $I = E/R$; this is known as Ohm's law. The reciprocal of resistance, G, is known as conductance; $G = 1/R$, so $I = EG$.

If a current of I amperes flows through a resistance of R ohms, the power dissipated is given as $W = I^2R$ watts (or joules per second). Resistance is often an unwanted property of conductors, as will appear later when we consider inductors. However, there are many applications where a resistor, a resistance of a known value, is useful. Wirewound

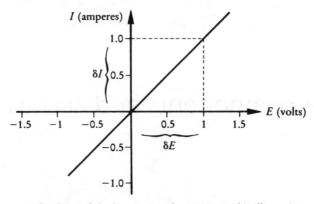

The slope of the line is given by δI/δE. In this illustration δI = 1 A and δE = 1 V, so the conductance G = 1 S. The S stands for siemens, the unit of conductance, formerly called the mho. G = 1/R.

Figure 1.1 *Current through a resistor of R ohms as a function of the applied voltage. The relation is linear, as shown, for a perfect resistor. At dc and low frequencies, most resistors are perfect for practical purposes.*

resistors use nichrome wire (high power types), constantan or manganin wire (precision types). They are available in values from a fraction of an ohm up to about a megohm, and can dissipate more power, size for size, than most other types but are mostly only suitable for use at lower frequencies, due to their self-inductance. For use at high frequencies, film or composition resistors are commonly used. Carbon film resistors are probably the commonest type used in the UK and Europe generally. They consist of a pyrolytically deposited film of carbon on a ceramic rod, with pressed-on end caps. Initially, the resistance is a few per cent of the final value: a spiral cut in the film is then made automatically, to raise the resistance to the designed value. Higher power or higher stability requirements are met by other resistor types using spiralled films of tin oxide or a refractory metal. The spiralling results in some self-inductance, which can be a disadvantage at radio frequencies; perhaps for this reason, carbon composition resistors are popular and widely used in the USA. These are constructed in a phenolic tube with lead-out wires inserted in the ends, and offer good RF performance combined with economy.

When two resistors are connected in series, the total resistance is the sum of the two resistances and when two resistors are connected in parallel, the total conductance is the sum of the two conductances. This is summarized in Figure 1.2. Variable resistors have three connections, one to each end of a resistive 'track' and one to the 'wiper' or 'slider'. The track may be linear or circular and adjustment is by screwdriver (preset

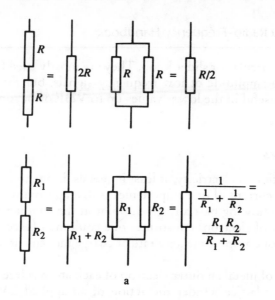

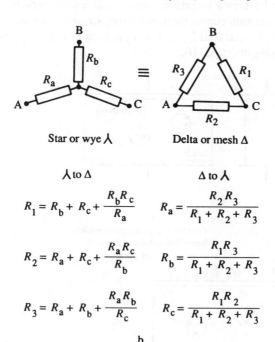

For resistors in series, total resistance is
$$R_t = R_1 + R_2 + R_3 \ldots$$

For resistors in parallel,
$$\frac{1}{R_t} = \frac{1}{R_1} + \frac{1}{R_2} + \frac{1}{R_3} \ldots$$

Star or wye $\curlywedge$ $\equiv$ Delta or mesh Δ

$\curlywedge$ to Δ

$$R_1 = R_b + R_c + \frac{R_b R_c}{R_a}$$

$$R_2 = R_a + R_c + \frac{R_a R_c}{R_b}$$

$$R_3 = R_a + R_b + \frac{R_a R_b}{R_c}$$

Δ to $\curlywedge$

$$R_a = \frac{R_2 R_3}{R_1 + R_2 + R_3}$$

$$R_b = \frac{R_1 R_3}{R_1 + R_2 + R_3}$$

$$R_c = \frac{R_1 R_2}{R_1 + R_2 + R_3}$$

Figure 1.2 *Resistors in combination.*
a Series parallel (also works for impedances).
b The star–delta transformation (also works for impedances, enabling negative values of resistance effectively to be produced).

types) or by circular or slider knob. They are mostly used for adjusting dc levels or the amplitude of low frequency signals, but the smaller preset sort can be useful in the lower values up to VHF or beyond.

Capacitors

The conduction of electricity, at least in metals, is due to the movement of electrons. A current of one ampere means that approximately 6242×10^{14} electrons are flowing past any given point in the conductor each second. This number of electrons constitutes one coulomb of electrical *charge*, so a current of one ampere means a rate of charge movement of one coulomb per second.

In a piece of metal an outer electron of each atom is free to move about in the atomic lattice. Under the action of an applied EMF, e.g. from a battery, electrons flow through the conductors forming the circuit, towards the positive terminal of the battery (i.e. in the opposite sense to the 'conventional' flow of current), to be replaced by other electrons flowing from the battery's negative terminal. If a capacitor forms part of the circuit, a continuous current cannot flow, since a capacitor consists of two plates of metal separated by a non-conducting medium, an insulator or a vacuum (see Figure 1.3a, b).

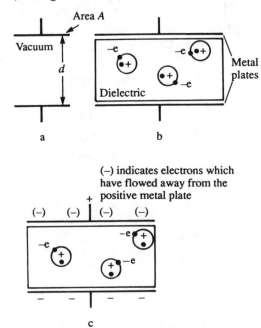

Figure 1.3 *Capacitors.*

A battery connected across the plates causes some electrons to leave the plate connected to its positive terminal, and an equal number to flow onto the negative plate (Figure 1.3c). A capacitor is said to have a *capacitance* C of one farad (1 F) if an applied EMF of one volt stores one coulomb (1 C) of charge. The capacitance is proportional to A, the area of the plates, and inversely proportional to their separation d, so that $C = k(A/d)$ (provided that d is much smaller than A). *In vacuo*, the value of the constant k is 8.85×10^{-12}, and it is known as the *permittivity of free space*, ε_0. Thus, *in vacuo*, $C = \varepsilon_0(A/d)$. More commonly, the plates of a capacitor are separated by air or an insulating solid substance; the permittivity of air is for practical purposes the same as that of free space. An insulator or *dielectric* is a substance such as air, polystyrene, ceramic, etc., which does not conduct electricity. This is because in an insulator all of the electrons are closely bound to the atoms of which they form part and cannot be completely detached except by an electrical force so great as to rupture and damage the dielectric. However, they can and do 'give' a little (Figure 1.3c), the amount being directly proportional to the applied voltage. This net displacement of charge in the dielectric enables a larger charge to be stored by the capacitor at a given voltage than if the plates were *in vacuo*. The ratio by which the stored charge is increased is known as the *relative permittivity*, ε_r. Thus $C = \varepsilon_0\varepsilon_r(A/d)$, and the stored charge $Q = CV$. Electronic circuits use capacitors as large as 500 000 μF (1 μF = 10^{-6} F), down to as small as 1 pF (one picofarad, 10^{-12} F), whilst stray capacitance of even a fraction of 1 pF can easily cause problems in RF circuits. On the other hand, very large electrolytic capacitors are used to store and smooth out energy in dc power supplies. The amount of energy J joules that a capacitor can store is given by $J = \frac{1}{2}CV^2$. (One joule of energy supplied every second represents a power of one watt.)

Although dc cannot flow through a capacitor, if a voltage of one polarity and then of the opposite polarity is repeatedly applied to a capacitor, charging current will always be flowing one way or the other. Thus an alternating EMF will cause a current to apparently flow through a capacitor. At every instant, $Q = CV$, so the greater the rate of change of voltage across the plates of the capacitor, the greater the rate of change of charge, i.e. the greater the current. If we apply a sinusoidal voltage $V = E_{max} \sin(\omega t)$* to a capacitor of CF, $Q = CE_{max} \sin(\omega t)$. The charge is a maximum at the peak of the voltage waveform, but at that instant the voltage (and the charge) is momentarily not changing, so the current is zero. It will have been flowing into the capacitor since the previous negative peak of the voltage, being a maximum where the rate of change

*ω is the 'angular velocity' in radians per second. There are 2π radians in a complete circle or cycle, so (for example) $\sin(20\pi t)$ would be a sinewave of ten cycles per second or 10 Hz, t indicating elapsed time in seconds.

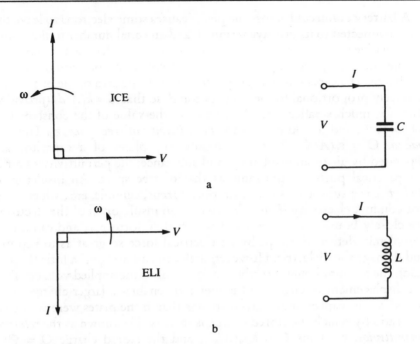

Figure 1.4 *Phase of voltage and current in reactive components.*
a ICE: the current I leads the applied EMF E (here V) in a capacitor.
The origin 0 represents zero volts, often referred to as ground.
b ELI: the applied EMF E (here V) across an inductor L leads the
current I.

of voltage was greatest, as it passed through zero. So the current is given
by $I = C \, \mathrm{d}v/\mathrm{d}t = \mathrm{d}(CE_{\max} \sin(\omega t))/\mathrm{d}t = \omega CE_{\max} \cos(\omega t)$. This means that
in a capacitor, the phase of the current leads that of the voltage by 90° (see
Figure 1.4). You can also see that, for a given $E_{\max}$, the current is proportional to the frequency of the applied alternating voltage. The 'reactance',
X_c, of a capacitor determines how much current flows for a given applied
alternating voltage E of frequency f (in hertz) thus: $I = E/X_c$, where $X_c =$
$1/(2\pi fC) = 1/(\omega C)$. X_c has units of ohms and we can take the 90° phase
shift into account by writing $X_c = 1/(j\omega C) = -j/(\omega C)$, where the
'operator' j indicates a +90° phase shift of the voltage relative to the
current. ($j^2 = -1$, so that $1/j = -j$). The $-j$ indicates a $-90°$ phase shift of the
voltage relative to the current, as in Figure 1.4. The reciprocal of reactance,
B, is known as susceptance; for a capacitor, $B = I/X_c = j\omega C$.

In addition to large electrolytics for smoothing and energy, already
mentioned, smaller sizes are used for 'decoupling' purposes, to bypass
unwanted ac signals to ground. At higher frequencies, capacitors using a
ceramic dielectric will often be used instead or as well, since they have
lower self-inductance. Small value ceramic capacitors can have a low

(nominally zero) temperature coefficient ('tempco'), using an NP0* grade of dielectric; values larger than about 220 pF have a negative temperature coefficient and for the largest value ceramic capacitors (used only for decoupling purposes), tempco may be as high as –15 000 parts per million per degree Celsius. Note that it is inadvisable to use two decoupling capacitors of the same value in parallel. Many other dielectrics are available, polystyrene being particularly useful as its negative tempco cancels (approximately) the positive tempco of ferrite pot inductor cores. Variable capacitors are used for tuned circuits, being either 'front panel' (user) controls, or preset types.

Inductors and transformers

A magnetic field surrounds any flow of current, such as in a wire or indeed a stroke of lightning. The field is conventionally represented by lines of magnetic force surrounding the wire, more closely packed near the wire where the field is strongest (Figure 1.5a and b) which illustrates the 'corkscrew rule' – the direction of the flux is clockwise viewed along the flow of the current. In Figure 1.5c, the wire has been bent into a loop: note that the flux lines all pass through the loop in the same direction. With many loops or 'turns' (Figure 1.5d) most of the flux encircles the whole 'solenoid': if there are N turns and the current is I amperes, then F, the magnetomotive force (MMF, analogous to EMF), is given by $F = NI$ amperes (sometimes called ampere turns). The resultant magnetic flux (analogous to current) is not uniform; it is concentrated inside the solenoid but spreads out widely outside as shown. If a long thin solenoid is bent into a loop or 'toroid' (Figure 1.5e) then all of the flux is contained within the winding and is uniform. The strength of the magnetic field H within the toroid depends upon the MMF per unit length causing it. In fact $H = I/l$ amperes/metre, where l is the length of the toroid's mean circumference and I is the effective current — the current per turn times the number of turns. The uniform magnetic field causes a uniform magnetic flux density, B webers/m², within the toroidal winding. The ratio B/H is called the *permeability of free space* μ_0, and its value is $4\pi \times 10^{-7}$. If the cross-sectional area of the toroid is A m², the total magnetic flux ϕ webers is $\phi = BA$. If the toroid is wound upon a ferromagnetic core, the flux for a given field strength is increased by a factor μ_r, the relative permeability. Thus $B = \mu_0\mu_r H$. Stated more fully, $\phi/A = \mu_0\mu_r F/l$ so that:

$$\phi = \frac{F}{l/(\mu_0\mu_r A)}$$

*N750 indicates a tempco of capacitance of –750 parts per million per °C: NP0 indicates a nominally zero tempco.

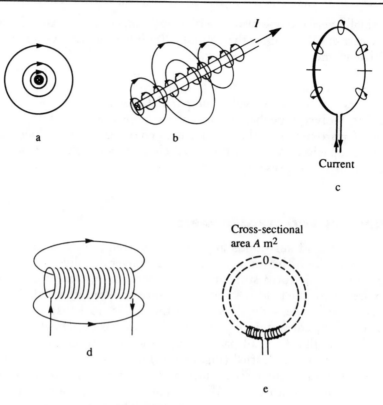

Figure 1.5 *The magnetic field.*
a End view of a conductor. The cross indicates current flowing into the paper (a point indicates flow out). By convention, the lines of flux surrounding the conductor are as shown, namely clockwise viewed in the direction of current flow (the corkscrew rule).
b The flux density is greatest near the conductor; note that the lines form complete loops, the path length of a loop being greater the further from the wire.
c Doughnut-shaped (toroidal) field around a single-turn coil.
d A long thin solenoid produces a 'tubular doughnut', of constant flux density within the central part of the coil.
e A toroidal winding has no external field. The flux density B within the tube is uniform over area A at all points around the toroid.

The term $l/(\mu_0\mu_r A)$ is called the *reluctance S* of the magnetic circuit, with units of amperes/weber, and is analogous to the resistance of an electric circuit. The magnetic circuit of the toroid in Figure 1.5e is uniform. If it were non-uniform, e.g. if there were a semicircular ferromagnetic core in the toroid extending half-way round, the total reluctance would simply be the sum of the reluctances of the different parts of the magnetic circuit, just as the total resistance of an electric circuit is the sum of all the parts in series.

When the magnetic field linking with a circuit changes, a voltage is induced in that circuit — the principle of the dynamo. This still applies, even if the flux is due to the current in that same circuit. An EMF applied to a coil will cause a current and hence a flux: the increasing flux induces an EMF in the coil in opposition to the applied EMF; this is known as Lenz's law. If the flux increases at a rate $d\phi/dt$, then the *back EMF* induced in each turn is $E_B = -d\phi/dt$, or $E_{Btotal} = -N\, d\phi/dt$ for an N turn coil. However,

$$\phi = \text{MMF/reluctance} = NI/S$$

and as this is true independent of time, their rates of change must also be equal:

$$d\phi/dt = (1/S)(dNI/dt)$$

So

$$E_{Btotal} = -N\, d\phi/dt = -N(1/S)(dNI/dt) = -(N^2/S)(dI/dt)$$

The term N^2/S, which determines the induced voltage resulting from unit rate of change of current, is called the *inductance L* and is measured in henrys:

$$L = N^2/S \text{ henrys}$$

If an EMF E is connected across a resistor R, a constant current $I = E/R$ flows. This establishes a potential difference (pd) V across the resistor, equal to the applied EMF, and the supplied energy I^2R is all dissipated as heat in the resistor. However, if an EMF E is connected across an inductor L, an increasing current flows. This establishes a back EMF V across the inductor (very nearly) equal to the applied EMF, and the supplied energy is all stored in the magnetic field associated with the inductor. At any instant, when the current is I, the stored energy is $J = \frac{1}{2}LI^2$ joules.

If a sinusoidal alternating current I flows through an inductor, a sinusoidal back EMF E_B will be generated. For a given current, as the rate of change is proportional to frequency, the back EMF will be greater, the higher the frequency. So the back EMF is given by

$$E_B = L\, dI/dt = L\, d(I_{max}\sin(\omega t))/dt = \omega L I_{max}\cos(\omega t)$$

This means that in an inductor, the phase of the voltage leads that of the current by 90° (see Figure 1.4). The 'reactance', X_L, of an inductor determines how much current flows for a given applied alternating voltage E of frequency f Hz thus: $I = E/X_L$, where $X_L = 2\pi fL = \omega L$. We can take the 90° phase advance of the voltage on the current into account by writing $X_L = j\omega L$. The reciprocal of reactance, B, is known as susceptance; for an

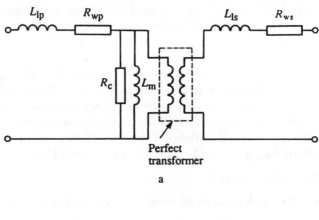

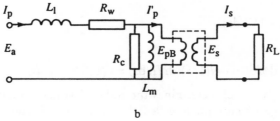

Figure 1.6 *Transformers.*
 a Full equivalent circuit.
 b Simplified equivalent circuit of transformer on load.

inductor, $B = 1/X_L = -j/\omega L$. Note that inductance is a property associated with the flow of current, i.e. with a complete circuit; it is thus meaningless to ask what is the inductance of a centimetre of wire in isolation. Nevertheless, it is salutary to remember (when working at VHF or above) that a lead length of 1 cm on a component will add an inductive reactance of about 6 Ω to the circuit at 100 MHz.

In practice, the winding of an inductor has a finite resistance. At high frequencies, this will be higher than the dc resistance, due to the 'skin effect', which tends to restrict the flow of current to the surface of the wire, reducing its effective cross-sectional area. The effective resistance is thus an increasing function of frequency. In some applications, this resistance is no disadvantage — it is even an advantage. An RF choke is often used in series with the dc supply to an amplifier stage, as part of the decoupling arrangements. The choke should offer a high impedance at RF, to prevent signals being coupled into/out of the stage, from or into other stages. The impedance should be high not only over all of the amplifier's operating frequency range, but ideally also at harmonics of the operating frequency (especially in the case of a class C amplifier) and way below the lowest operating frequency as well, since there the gain of RF power transistor is often much greater. A sectionalized choke, or two chokes of very different values in series may be required. At UHF, an effective ploy is the graded choke, which is close wound at one end but progressively pulled out to wide spacing at the other. It should be wound with the thinnest wire which will carry the required dc supply current and can with advantage be wound with resistance wire. A very effective alternative at VHF and UHF is to slip a ferrite bead or two over a supply lead. They are available in a grade of ferrite which becomes very lossy above 10 MHz so that at RF there is effectively a resistance in series with the wire, but with no corresponding loss at dc. Where an inductor is to form part of a tuned circuit on the other hand, one frequently requires the lowest loss resistance (highest Q) possible. At lower RF frequencies, up to a few megahertz, gapped ferrite pot cores (inductor cores) are very convenient, offering a Q which may be as high as 900. The best Q is obtained with a single layer winding. The usual form of inductor at higher frequencies, e.g. VHF, is a short single-layer solenoid, often fitted with a ferrite or dust iron slug for tuning and sometimes with an outer ferromagnetic hood and/or metal can for screening. A winding spaced half a wire diameter between turns gives a 10 to 30% higher Q than a close spaced winding.

Two windings on a common core form a 'transformer', permitting a source to supply ac energy to a load with no direct connection, Figure 1.6. Performance is limited by core and winding losses and by leakage inductance, as covered more fully in Chapter 3.

Passive circuits

Resistors, capacitors and inductors can be combined for various purposes. When a circuit contains both resistance and reactance, it presents an 'impedance' Z which varies with frequency. Thus $Z = R + j\omega L$ (resistor in series with an inductor) or $Z = R - j/(\omega C)$ (resistor in series with a capacitor). The reciprocal of impedance, Y, is known as admittance:

$$Y = 1/Z = S - j/\omega L \qquad \text{or} \qquad Y = S + j\omega C$$

At a given frequency, a resistance and a reactance in series R_s and X_s behaves exactly like a different resistance and reactance in parallel R_p and X_p. Occasionally, it may be necessary to calculate the values of R_s and X_s given R_p and X_p, or vice versa. The necessary formulae are given in Appendix 1.

Since the reactance of an inductor rises with increasing frequency, that of a capacitor falls, whilst the resistance of a resistor is independent of frequency, the behaviour of the combination will in general be frequency dependent. Figure 1.7 illustrates the behaviour of a series resistor–shunt capacitor

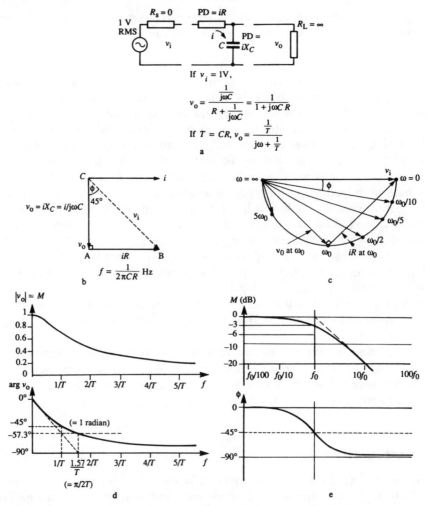

Figure 1.7 CR *low-pass (top cut) lag circuit (see text).*

(low pass) combination. Since the current through a capacitor leads the voltage across it by 90°, at that frequency (ω_0) where the reactance of the capacitor in ohms equals the value of the resistor, the voltage and current relationships in the circuit are as in Figure 1.7b. The relation between v_i and v_o at ω_0 and other frequencies is shown in the 'circle diagram' (Figure 1.7c). Figure 1.7d plots the magnitude or modulus M and the phase or argument ϕ of v_o versus a linear scale of frequency, for a fixed v_i. Note that it looks quite different from the same thing plotted to the more usual logarithmic frequency scale (Figure 1.7e).

If C and R in Figure 1.7a are interchanged, a high-pass circuit results, whilst low- and high-pass circuits can also be realized with a resistor and an inductor. All the possibilities are summarized in Figure 1.8. Figure 1.9a

Curve no.	Voltage output into open circuit	Current output into short circuit	Voltage output into open circuit	Current output into short circuit
1	$\dfrac{j\omega T}{1+j\omega T}$	$\dfrac{1}{R}\cdot\dfrac{j\omega T}{1+j\omega T}$	$R\,\dfrac{j\omega T}{1+j\omega T}$	$\dfrac{j\omega T}{1+j\omega T}$
2	$\dfrac{1}{1+j\omega T}$	$\dfrac{1}{R}\,\dfrac{1}{1+j\omega T}$	$R\,\dfrac{1}{1+j\omega T}$	$\dfrac{1}{1+j\omega T}$
3		$\dfrac{1}{R}\cdot\dfrac{1+j\omega T}{j\omega T}$	$R\,\dfrac{1+j\omega T}{j\omega T}$	
4		$\dfrac{1}{R}(1+j\omega T)$	$R(1+j\omega T)$	
5		$j\omega C$	$j\omega L$	
6		$\dfrac{1}{j\omega L}$	$\dfrac{1}{j\omega C}$	

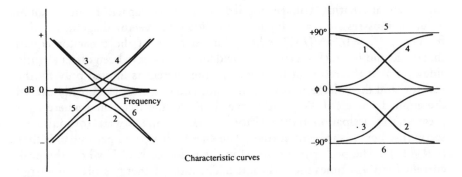

Characteristic curves

Figure 1.8 *All combinations of one resistance and one reactance, and of one reactance only, and their frequency characteristics (magnitude and phase) and transfer functions (reproduced by courtesy of Electronics and Wireless World).*

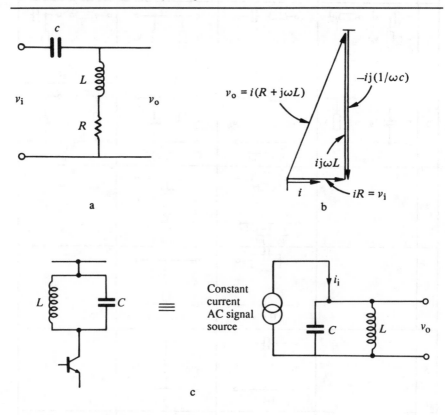

Figure 1.9 *Series and shunt-fed tuned circuits.*
a Series resonant tuned circuit.
b Vector diagram of same at f_r.
c Shunt current fed parallel tuned circuit.

shows an alternating voltage applied to a series capacitor and a shunt inductor-plus-resistor, and Figure 1.9b shows the vector diagram for that frequency ($f_r = 1/2\pi\sqrt{[LC]}$) where the reactance of the capacitor equals that of the inductor. (For clarity, coincident vectors have been offset slightly sideways.) At the resonant frequency f_r, the current is limited only by the resistor, and the voltage across the inductor and capacitor can greatly exceed the applied voltage if X_L greatly exceeds R. At the frequency where v_o is greatest, the dissipation in the resistor is a maximum, i^2R watts (or joules per second), where i is the rms current. The energy dissipated per radian is thus $(i^2R)/(2\pi f)$. The peak energy stored in the inductor is $\frac{1}{2}LI^2$ where the peak current I is 1.414 times the rms value i. The ratio of energy stored to energy dissipated per radian is thus $(\frac{1}{2}L(\sqrt{2}i)^2)/\{(i^2R)/(2\pi f)\} = 2\pi fL/R = X_L/R$, the ratio of the reactance of the inductor (or of the capacitor) at resonance to the resistance. If there is no separate resistor, but R represents simply the effective resistance of the winding of the inductor at frequency f, then the ratio is known as the Q (quality factor) of the inductor at that frequency. Capacitors also have effective series resistance, but it tends to be very much lower than for an inductor: they have a much higher Q. So in this case, the Q of the tuned circuit is simply equal to that of the inductor. Figure 1.9c shows a parallel tuned circuit, fed from a very high source resistance, a 'constant current generator'. The response is very similar to that shown in Figure 1.9b for the series tuned circuit, especially if Q is high. However, maximum v_o will not quite occur when it is in phase with v_i unless the Q of the inductor equals that of the capacitor.

A tuned circuit passes a particular frequency or band of frequencies, the exact response depending upon the Q of the circuit. Relative to the peak, the -3 dB bandwidth δf is given by $\delta f = f_0/Q$, where f_0 is the resonant frequency (see Appendix 4). Where greater selectivity is required than can be obtained from a single tuned circuit, two options are open. Subsequent tuned circuits can be incorporated at later stages in, e.g. a receiver: they may all be tuned to exactly the same frequency ('synchronously tuned'), or if a flatter response over a narrow band of frequencies is required, they can be slightly offset from each other ('stagger tuned'). Alternatively, two tuned circuits may be coupled together to provide a 'band-pass' response. At increasing offsets from the tuned frequency, they will provide a more rapid increase in attenuation than a single tuned circuit, yet with proper design they will give a flatter pass band. The flattest pass band is obtained with critical coupling; if the coupling is greater than this, the pass band will become double-humped, with a dip in between the peaks. Where the coupling between the two tuned circuits is by means of their mutual inductance M, the coefficient of coupling k is given by

$$k = M/\sqrt{(L_p L_s)} = M/L$$

if the inductance of the primary tuned circuit equals that of the secondary. The value of k for critical coupling

$$k_c = 1/\sqrt{(Q_p Q_s)} = 1/Q$$

if the Q of the primary and secondary tuned circuits is equal. Thus for example, if $Q_p = Q_s = 100$ then

$$k_c = 0.01 = M/L, \qquad \text{if } L_p = L_s$$

So just 1% of the primary flux should link the secondary circuit. Many other types of coupling are possible, some of which are shown in Figure 1.10; Terman [1] gives expressions for the coupling coefficients for these and other types of coupling circuits.

Where a band-pass circuit is tunable by means of ganged capacitors C_p and C_s (Figure 1.10a and b), the coupling will vary across the band. A judicious combination of top and bottom capacitive coupling can give a nearly constant degree of coupling across the band. To this end, the coupling

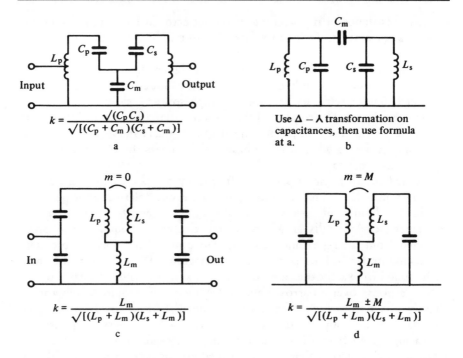

Figure 1.10 *Coupled tuned circuits.*
a Bottom capacitance coupling.
b Top capacitance coupling.
c Bottom inductive coupling.
d Mixed mutual and bottom inductive coupling.

capacitors C_m may be trimmers to permit adjustment on production test. Where C_m in Figure 1.10b turns out to need an embarrassingly small value of trimmer, two small fixed capacitors of 1 pF or so in series may be used, with a much larger trimmer from their junction to ground.

Figure 1.7 showed a simple low-pass circuit. Its final rate of attenuation is only 6 dB/octave and the transition from the pass band to the stop band is not at all sharp. Where a sharper transition is required, a series L in place of the series R offers a better performance. If R_L = infinity, R_s = 1.414X_L at ω_0 (where $\omega_0 = 1\sqrt{LC}$), the attenuation is 3 dB at ω_0, flat below that frequency and tends to –12 dB/octave above it. If a little peaking in the passband is acceptable ($R_s = X_L$ at ω_0), there is no attenuation at all at ω_0 and the cut-off rate settles down soon after to 12 dB/octave as before. This is an example of a second order Chebychev response. To get an even faster rate of cut-off, especially if we require a flat pass band with no peaking (a Butterworth response), we need a higher order filter. Figure 1.11a shows

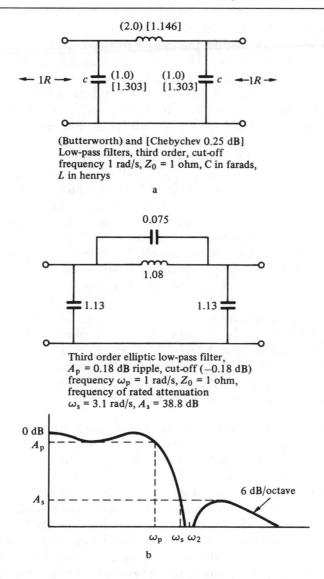

(2.0) [1.146]

← 1R →   c   (1.0)   (1.0)   c   ← 1R →
              [1.303]   [1.303]

(Butterworth) and [Chebychev 0.25 dB]
Low-pass filters, third order, cut-off
frequency 1 rad/s, Z_0 = 1 ohm, C in farads,
L in henrys

a

0.075

1.08

1.13   1.13

Third order elliptic low-pass filter,
A_p = 0.18 dB ripple, cut-off (−0.18 dB)
frequency ω_p = 1 rad/s, Z_0 = 1 ohm,
frequency of rated attenuation
ω_s = 3.1 rad/s, A_s = 38.8 dB

0 dB
A_p

6 dB/octave

A_s

ω_p ω_s ω_2

b

Figure 1.11   *Butterworth, Chebychev and Elliptic three-pole low-pass filter.*

a third order filter designed to work from a 1 Ω source into a 1 Ω load,
with a cut-off frequency of 1 rad/s, i.e. $1/2\pi$ = 0.159 Hz. (These
'normalized' values are not very useful as they stand, but to get to, say, a
2 MHz cut-off frequency, simply divide all the component values by 2 ×
10^6, and to get to a 50 Ω design divide all the capacitance values by 50 and

multiply all the inductance values by 50. Thus starting with normalized values you can easily modify the design to any cut-off frequency and impedance level you want.) The values in round brackets are for a Butterworth design and those in square brackets for a 0.25 dB Chebychev design, i.e. one with a 0.25 dB dip in the pass band. Note the different way that Butterworth and Chebychev filters are specified: the values shown will give an attenuation at 0.159 Hz of 3 dB for the Butterworth filter, but a value equal to the pass-band ripple depth (–0.25 dB for the example shown) for the Chebychev filter. Even so, the higher order Chebychev types, especially those with large ripples, will still show more attenuation in the stop band than Butterworth types. Both of the filters in Figure 1.11a cut off at the same ultimate rate of 18 dB/octave. However, if they were designed for the same –3 dB frequency, the Chebychev response would show much more attenuation at frequencies well into the stop band, because of its steeper initial rate of cut-off, due to the peaking. Most of the filter types required by the practising RF engineer can be designed with the use of published normalized tables of filter responses [2, 3]. These also cover elliptic filters, which offer an even faster descent into the stop band, if you can accept a limitation on the maximum attenuation as shown in Figure 1.11b. For details of more specialized filters such as helical resonator or combline band-pass filters, mechanical, ceramic, quartz crystal and SAW filters, etc., the reader should refer to one of the many excellent books dealing specifically with filter technology. However, the basic quartz crystal resonator is too important a device to pass over in silence.

A quartz crystal resonator consists of a ground, lapped and polished crystal blank upon which metallized areas (electrodes) have been deposited. There are many different 'cuts' but one of the commonest, used for crystals operating in the range 1 to 200 MHz is the AT cut, used both without temperature control and, for an oscillator with higher frequency accuracy, in an oven maintained at a constant temperature such as +70°C, well above the expected top ambient temperature (an OCXO). Where greater frequency accuracy than can be obtained with a crystal at ambient temperature is required, but the warm-up time or power requirements of an oven are unacceptable, a temperature-compensated crystal oscillator (TCXO) can be used. Here, temperature-sensitive components such as thermistors are used to vary the reverse bias on a voltage-variable capacitor in such a way as to reduce the dependence of the crystal oscillator's frequency upon temperature.

When an alternating voltage is applied to the crystal's electrodes, the voltage stress in the body of the quartz (which is a very good insulator) causes a minute change in dimensions, due to the piezo-electric effect. If the frequency of the alternating voltage coincides with the natural frequency of vibration of the quartz blank, which depends upon its size and thickness and the area of the electrodes, the resultant mechanical vibrations are much greater than otherwise. The quartz resonator behaves in fact like a series tuned circuit, having a very high L/C ratio. Despite this, it still displays a very low ESR (equivalent series resistance) at resonance, due to its very high effective Q, typically in the range of 10 000 to 100 000. Like any series tuned circuit, it appears inductive at frequencies above resonance and there is a frequency at which this net inductance resonates with C_0, the capacitance between the electrodes. Since even for a crystal operating in the MHz range, L may be several henrys and C around a hundredth of one picofarad, the difference between the resonant (series resonant) and the antiresonant (parallel resonance with C_0) frequencies may be less than 0.1% (see Appendix 9). A crystal may be specified for operation at series or at parallel resonance and the manufacturer will have adjusted it appropriately to resonate at the specified frequency. Crystals operating at frequencies below about 20 MHz are usually made for operation at parallel resonance, and operated with 30 pF of external circuit capacitance C_c in parallel with C_0. Trimming C_c allows for final adjustment of the operating frequency in use. This way, a crystal's operating frequency may be 'pulled', perhaps by as much as one or two hundred parts per million, but the more it is pulled from its designed operating capacitance, the worse the frequency stability is likely to be. Like many mechanical resonators (e.g. violin string, brass instrument), a crystal can vibrate at various harmonics or overtones. Crystals designed for use at frequencies much above 20 MHz generally operate at an overtone such as the 3rd, 5th, 7th or 9th. These are generally operated at or near series resonance. Connecting an adjustable inductive or capacitive reactance, not too large compared to the ESR, in series permits final adjustment to frequency in the operating circuit, but the pulling range available with series operation is not nearly as great as with parallel operation. The greatest frequency accuracy is obtained from crystals using the 'SC' (strain compensated or doubly rotated) cut, although these are considerably more expensive. They are also slightly more difficult to apply, as they have more spurious resonance modes than AT cut crystals, and these have to be suppressed to guarantee operation at the desired frequency.

Quartz crystals are also used in band-pass filters, where their very high Q permits very selective filters with a much smaller percentage bandwidth to be realized than would be possible with inductors and capacitors. Traditionally, the various crystals, each pretuned to its designed frequency, were coupled together by capacitors in a ladder or lattice circuit. More recently, pairs of crystals ('monolithic dual resonators') are made on a single blank, the coupling being by the mechanical vibrations. More recently still, monolithic quad resonators have been developed, permitting the manufacture of smaller, cheaper filters of advanced performance.

References

1. Terman, F. E. *Radio Engineers' Handbook*, McGraw-Hill (1943)
2. Zverev, A. I. *Handbook of Filter Synthesis*, John Wiley & Sons (1967)
3. Geffe, P. R. *Simplified Modern Filter Design*, Iliffe (1964)

2

RF transmission lines

RF transmission lines are used to convey a radio frequency signal with minimum attenuation and distortion. They are of two main types, balanced and unbalanced. A typical example of the former is the flat twin antenna feeder with a characteristic impedance of 300 Ω often used for VHF broadcast receivers, and of the latter is the low loss 75 Ω coaxial downlead commonly used between a UHF TV set and its antenna. Characteristic impedance can be explained in conjunction with Figure 2.1 as follows. Leaving aside the theoretical ideal voltage source, any practical generator (source of electrical power, e.g. a battery) has an associated internal resistance, and the maximum power that can be obtained from it flows in a load whose resistance equals the internal resistance. In the case of a source of RF energy, for example a signal generator, it is convenient if the source impedance is purely resistive, i.e. non-reactive, as then the power delivered to a resistive load (no power can ever be delivered to a purely reactive load) will be independent of frequency. In Figure 2.1a and b, a source resistance of 1 Ω and a maximum available power of 1 W is shown, for simplicity of illustration. However, the usual source resistance for a signal generator is 50 Ω unbalanced, that is to say the output voltage appears on the inner lead of a coaxial connector whose outer is earthy (carries no potential with respect to ground). Imagine such an output connected to an infinitely-long loss-free coaxial cable. If the diameters of the inner and outer conductors are correctly proportioned (taking into account the permittivity of the dielectric), the signal generator will deliver the maximum energy possible to the cable; the cable will appear to the source as a 50 Ω load and the situation is the same as if a 50 Ω resistor terminated a finite length of the cable. Figure 2.1c shows a short length of a balanced feeder, showing the series resistance and inductance of the conductors and the parallel capacitance and conductance between them, per unit length (the conductance is usually negligible). Denoting the series and parallel impedances as Z_s and Z_p respectively, the characteristic impedance Z_0 of the line is given by $Z_0 = \sqrt{(Z_s Z_p)}$. If G is

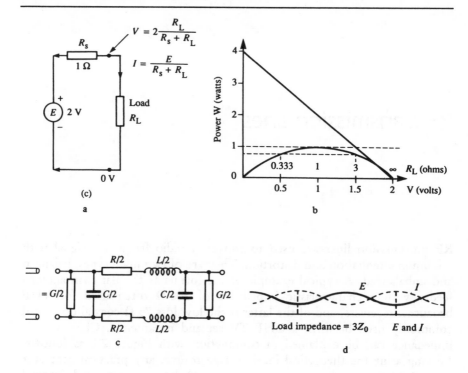

Figure 2.1 *Matching and transmission lines.*

a Source connected to a load R$_L$.

b E = 2V, R$_S$ = 1 Ω. Maximum power in the load occurs when R$_L$ = R$_S$ and V = E/2 (the matched condition, but only falls by 25% for R$_L$ = 3R$_S$ and R$_L$ = R$_S$/3. For the matched case the total power supplied by the battery is twice the power supplied to the load. On short-circuit, four times the matched load power is supplied, all dissipated internally in the battery.

c Two-wire line: balanced π equivalent of short section.

d Resultant voltage and current standby waves when load resistance = 3Z$_0$.

negligible and $j\omega L \gg R$, then practically $Z_0 = \sqrt{(L/C)}$ and the phase shift β along the line is $\sqrt{(LC)}$ radians per unit length. Thus the wavelength of the signal in the line (always less than the wavelength in free space) is given by $\lambda = 2\pi/\beta$. Although at RF, $j\omega L \gg R$, the resistance is still responsible for some losses, so that the signal is attenuated to some extent in its passage along the line. The attenuation per unit length is given by the full expression for the propagation constant $\gamma = \alpha + j\beta = \sqrt{(Z_s/Z_p)} = \sqrt{\{(R + j\omega L)(G + j\omega C)\}}$ where α is the attenuation constant per unit length, in nepers. Nepers express a power ratio in terms of natural logs, i.e. to base e rather than to base 10: 1 neper = 8.7 dB. In practice, R will be greater than the dc resistance, due to the skin effect, which increases with frequency; the attenuation 'constant' is therefore not really a constant, but increases with increasing frequency.

If a 50 Ω source feeds a lossless 50 Ω coaxial cable but the load at the far end of the cable is higher or lower than 50 Ω, then the voltage appearing across the load will be higher or lower and the current through it lower or higher respectively than for a matched 50 Ω load. Some of the voltage incident upon the load is reflected back towards the source, either in phase or in antiphase, and this reflected wave travels back towards the source with the same velocity as the incident wave: this is illustrated in Figure 2.1d for the case of a 150 Ω load connected via a 50 Ω cable to a 50 Ω source, i.e. a load of 3 $\times$ Z_0. The magnitude of the reflected current relative to the incident current is called the reflection coefficient, ρ, and is given by

$$\rho = (Z_0 - Z_L)/(Z_0 + Z_L)$$

In Figure 2.1d, since $Z_L = 3Z_0$, $\rho = -0.5$, the minus sign indicating that the reflected current is reversed in phase. Thus if the incident voltage and current is unity, the net current in the load is the sum of the incident and reflected currents, $= 1 - 0.5 = 0.5$ A. The net voltage across the load is increased (or decreased) in the same proportion as the current is decreased (or increased), so the net voltage across the load is 150% and varies along the line between this value and 50% of the incident voltage. The ratio of the maximum to minimum voltage along the line is called the 'voltage standing wave ratio', VSWR, and is given by VSWR $= (1 - \rho)/(1 + \rho)$ (or its reciprocal, whichever is greater than unity), so for the case in Figure 2.1d where $\rho = -0.5$, the VSWR $= 3$. In a line terminated in a resistive load equal to the characteristic impedance Z_0 (a matched line), $\rho = 0$ and the VSWR equals unity.

If a length of 50 Ω line is exactly $\lambda/2$ or a whole number multiple thereof, the source in Figure 2.1d will see a 150 Ω load, but if it is $\lambda/4$, $3\lambda/4$, etc., it will see a load of 16.7 Ω. In fact, a quarter-wavelength of line acts as a transformer, transforming a resistance R_1 into a resistance R_2, where $R_1 \times R_2 = Z_0^2$. The same goes for reactances X_1 and X_2 (but note that if X_1 is capacitive X_2 will be inductive and vice versa) and for complex impedances Z_1 and Z_2. Thus a quarter-wavelength of line of characteristic impedance $\sqrt{(R_1 R_2)}$ can match a load R_2 to a source R_1 at one spot frequency, and over about a 10% bandwidth in practice. Note that the electrical length of a line depends upon the frequency in question. If a line is exactly $\lambda/4$ long at one frequency, it will appear shorter than $\lambda/4$ at lower frequencies and longer at higher, so a quarter-wave transformer is inherently a narrow band device. A quarter-wave transformer will transform a short circuit into an open circuit and vice versa, and a line less than $\lambda/4$ will transform either into a pure reactance. This is illustrated in Figure 2.2a. Power (implying current in phase with the voltage) is shown flowing along a loss-free RF cable towards an open circuit. (Figure 2.2a is a snapshot at a single moment in time; the vectors further along the line appear advanced since they will not reach the same phase as the input vectors until a little later on.) On arriving, no power can be dissipated as there is no resistance; the conditions must in fact be exactly the same as would apply at the output of the generator in Figure 2.1a if it were unterminated, i.e. an open-circuit terminal voltage of twice the voltage which would exist across a matched load, and no current flowing. The only way this condition can be met is if there is a reflected wave at the open-circuit end of the feeder, with its voltage in phase with the incident voltage and its current in antiphase with the incident current. This wave propagates back towards the source and Figure 2.2a also shows the resultant voltage and current. It can be seen that at a distance of $\lambda/8$ from the open circuit, the voltage is lagging the current by 90°, as in a capacitor. Moreover, the ratio of voltage to current is the same as for the incident wave, so the reactance of the apparent capacitance in ohms equals the characteristic impedance of the line. The reactance is less than this approaching $\lambda/4$ and greater approaching the open end of the line. Similarly, for a line less than $\lambda/4$ long, a short-circuit termination looks inductive.

The way impedance varies with line length for any type of termination is neatly represented by the Smith chart (Figure 2.2b). The centre of the chart represents Z_0, and this is conventionally shown as a 'normalized' value of

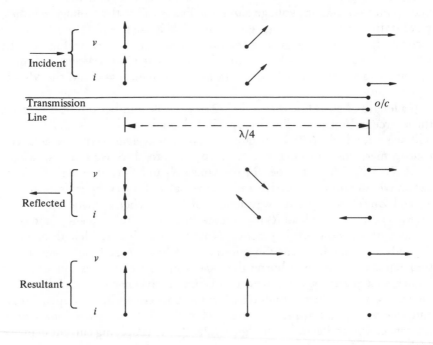

a

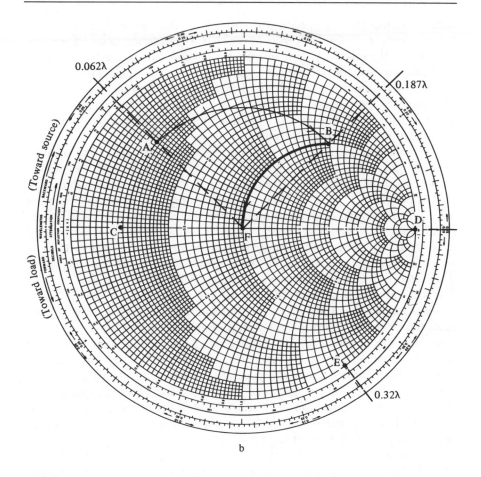

b

Figure 2.2
a At λ/8 from an open circuit, the current leads the voltage by 90°, i.e. at this point an o/c line looks like a capacitance C with a reactance of $1/jZ_0$. At λ/4, C = ∞.
b The Smith chart.

unity. To get to practical values, simply multiply all results by Z_0, e.g. by 50 for a 50 Ω system. The chart can be used equally well to represent impedances or admittances. The horizontal diameter represents all values of pure resistance or conductance, from zero at the left side to infinity at the right. Circles tangential to the right-hand side represent impedances with a constant series resistive component (or admittances with a constant shunt conductance component). Arcs branching leftwards from the right-hand side are loci of impedances (admittances) of constant reactance (susceptance), in the upper half of the chart representing inductive reactance or capacitive susceptance. Circles concentric with the centre of the chart are loci of constant VSWR, the centre of the chart representing unity VSWR and the edge of the chart a VSWR of infinity. Distance along the line from the load back towards the source can conveniently be shown clockwise around the periphery, one complete circuit of the chart equalling half a wavelength. The angle of the reflection coefficient, which is in general complex (only being a positive or negative real number for resistive loads) can also be shown around the edge of the chart.

The Smith chart can be used to design spot frequency matching arrangements for any given load, using lengths of transmission line. (It can also be used to design matching networks using lumped capacitance and inductance; see Appendix 1.) Thus in Figure 2.2b, using the chart to represent normalized admittances, the point A represents a conductance of 0.2 in parallel with a (capacitive) susceptance of $+j0.4$. Moving a distance of $(0.187 - 0.062)\lambda = 0.125\lambda$ towards the source brings us to point B where the admittance is conductance 1.0 in parallel with $+j2.0$ susceptance. (Continuing around the chart on a constant VSWR circle to point C tells us that without matching, the VSWR on the line would be $1/0.175 = 5.7$.) Just as series impedances add directly, so do shunt admittances. So if we add an susceptance of $-j2.0$ across the line at a point 0.125λ from the load, it will cancel out the susceptance of $+j2.0$ at point B. In fact, the inductive shunt susceptance of $-j2.0$ parallel resonates with the $+j2.0$ capacitive susceptance, so that viewed from the generator, point B is moved round the constant conductance line to point F, representing a perfect match. The $-j2.0$ shunt susceptance can be a 'stub', a short-circuit length of transmission line. Point E represents $-j2.0$ susceptance and the required length of line starting from the short circuit at D is $(0.32 - 0.25)\lambda = 0.07\lambda$. This example of matching using lengths of transmission lines ignores the effect of any losses in the lines. This is permissible in practice as the lengths involved are so small, but where longer runs (possibly many wavelengths) of coaxial feeder are involved, e.g. to or from an antenna, the attenuation may well be significant. It will be necessary to select a feeder with a low enough loss per unit length at the frequency of interest to be acceptable in the particular installation.

Matching using lengths of transmission line can be convenient at frequencies from about 400 MHz upwards. Below this frequency, things start to get unwieldy, and lumped components, inductors and capacitors, are thus usually preferred. In either case, the match is narrow band, typically holding reasonably well over a 10% bandwidth.

3
RF transformers

RF transformers are used for two main purposes: to convert from one impedance level to another, or to provide electrical isolation between two circuits. Often, of course, isolation and impedance conversion are both required, and a suitable transformer fulfills both these functions with minimal power loss. Examples of transformers used mainly for isolation include those used to couple in and out of data networks and pulse transformers for SCR firing. Examples used mainly for impedance conversion include interstage transformers in MOSFET VHF power amplifiers and the matching transformer between a 50 Ω feeder and a 600 Ω HF antenna. Such a matching transformer may also be required to match an unbalanced feeder to a balanced antenna. With so many basically different applications, it is no wonder that there is a wide range of transformer styles, from small-signal transformers covering a frequency range approaching 100 000:1, to high power HF transformers where it is difficult to cover more than a few octaves.

Before describing the techniques special to RF transformers, it may be helpful to recap on the operation of transformers in general. Transformer action depends upon as much as possible (ideally all) of the magnetic flux surrounding a primary winding linking with the turns of a secondary winding, to which end a core of high permeability magnetic material is often used (Figure 3.1a). Even so, some primary current — the magnetizing current — will be drawn, even when no secondary current flows: this magnetizing current causes the flux Φ, with which it is in phase. The alternating flux induces in the primary a back-EMF E_{pB} nearly equal to the applied voltage E_a (Figure 3.1b). The amount of magnetizing current drawn will depend upon the primary or magnetizing inductance L_m, which in turn depends upon the number of primary turns and the reluctance of the core: the reluctance depends upon the permeability of the core material and the dimensions. There will be some small power loss associated with the alternating flux on the core, due to hysteresis and eddy current losses in the

27

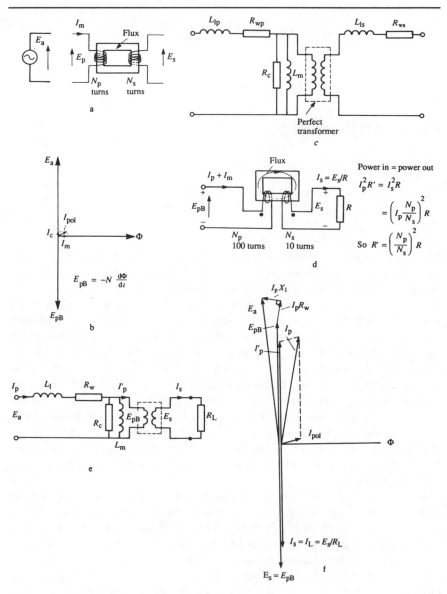

Figure 3.1 *Transformer operation (see text).*

core material. This can be represented by a core loss resistance R_c, connected (like the magnetizing inductance L_m) in parallel with the primary of a fictional ideal transformer (Figure 3.1c). The core loss resistance draws a small primary current I_c in phase with the applied voltage E_a, and this together with the quadrature magnetizing current I_m forms the primary off-load current I_{pol} (Figure 3.1b).

Figure 3.1d shows how (ignoring losses) a load resistance R connected to the secondary winding, appears at the transformer input as a resistance R' transformed in proportion to the square of the turns ratio. In practice, there are other minor imperfections to take into account as follows. Firstly, there will be a finite winding resistance R_{wp} associated with the primary winding, and similarly with the secondary winding. Also, not quite all of the flux due to I_m in the primary winding will link with the secondary winding; this is called the primary leakage inductance L_{lp}. If we were to apply E_a to the secondary winding, a similar effect would be observed and the secondary leakage inductance is denoted by L_{ls}. These are both shown, along with L_m and R_c, in Figure 3.1c. With negligible error usually, the secondary leakage inductance and winding resistance can be translated across to the primary (by multiplying them by the square of the turns ratio) and added to the corresponding primary quantities, to give an equivalent total leakage inductance and winding resistance L_l and R_w (Figure 3.1e). Figure 3.1f shows the transformer of Figure 3.1e on load, taking the turns ratio to be unity, for simplicity. For any other ratio, E_{pb}/E_s and I_s/I_p' would simply be equal to the turns ratio N_p/N_s. You can see that at full load, the total primary current is almost in antiphase with the secondary current, and that if the load connected to the secondary is a resistance (as in Figure 3.1e and f), then the primary current lags the applied voltage very slightly, due to the finite magnetizing current.

The foregoing analysis is perfectly adequate in the case of a mains power transformer, operating at a fixed frequency, but it is decidedly oversimplified in the case of a wideband signal transformer, since it ignores the self- and interwinding-capacitances of the primary and secondary. Unfortunately it is not easy to take these into account analytically, or even show them on the transformer circuit diagram, since they are distributed and cannot be accurately represented in a convenient lumped form like L_m, L_l, R_c and R_w. However, they substantially influence the performance of a wideband RF transformer at the upper end of its frequency range, particularly in the case of a high impedance winding, such as the secondary of a 50 Ω to 600 Ω transformer rated at kilowatts and matching an HF transmitter to a rhombic antenna, for instance. With certain assumptions, values for the primary self-capacitance and for the equivalent secondary self-capacitance referred to the primary can be calculated from formulae quoted in the literature [1]. This can assist in deciding whether in a particular design, the capacitance or the leakage inductance will have most effect in limiting the transformer's upper 3 dB point.

When developing a design for a wideband transformer, it is necessary to have some idea of the values of the various parameters in Figure 3.1e. In addition to calculation, as mentioned above concerning winding capacitances, two other approaches are possible: direct measurement and deduction. Direct measurement of L_m and L_l is straightforward and the results will be reasonably accurate if the measurement is performed near the lower end of the transformer's frequency range, where the effect of winding capacitance is minimal. The primary inductance is measured with the transformer off load, i.e. with the secondary open circuit. With the secondary short circuited on the other hand, a (near) short circuit will be reflected at the primary of the perfect transformer, so L_m and R_c will both be shorted out. The measurement therefore gives the total leakage inductance referred to the primary. The measured values of both primary and leakage inductance will exhibit an associated loss component, due to R_c and R_w respectively. In former times the measurements would have been made at spot frequencies using an RF bridge — a time consuming task. Nowadays, the open- and short-circuit primary impedances can be readily observed, as a function of frequency, as an s_{11} measurement on an s-parameter test set.

The second approach to parameter evaluation is by deduction from the performance of the transformer with its rated load connected. The primary inductance is easily determined since it will result in a 3 dB insertion loss, as the operating frequency is reduced, at that frequency where its reactance has fallen to the value of the rated nominal primary resistance and the source resistance in parallel, i.e. 25 Ω in a 50 Ω system. Note that the relevant frequency is not that at which the absolute insertion loss is 3 dB, but that at which it has increased by 3 dB relative to the midband insertion loss. Even this is a simplification, assuming as it does that the midband insertion loss is not influenced by L_l, and that R_w and R_c are constant with frequency, which is only approximately true. At the top end of the transformer's frequency range, things are more difficult, as the performance will be influenced by both the leakage inductance and the self- and interwinding-capacitances and by the core loss R_c. The latter may increase linearly with frequency, but often faster than this, especially in high-power transformers running at a high flux density. The relative importance of leakage inductance and stray capacitance in determining high frequency performance will depend upon the impedance level of the higher impedance winding, primary or secondary as the case may be. With a high ratio transformer, it may be beneficial to suffer some increase in leakage inductance in order to minimize the self-capacitance of the high impedance (e.g. 600 Ω) winding: in any case, in a high power RF transformer increased spacing of the secondary layer may be necessary to prevent danger of voltage breakdown in the event of an open circuit, such as an antenna fault.

In low-power (and hence physically small) transformers of modest ratio, leakage inductance will usually be more of a problem than self-capacitance. Here, measures can be taken to maximize the coupling between primary and secondary. Clearly, the higher the permeability of the core material used, the less turns will be necessary to achieve adequate primary inductance.

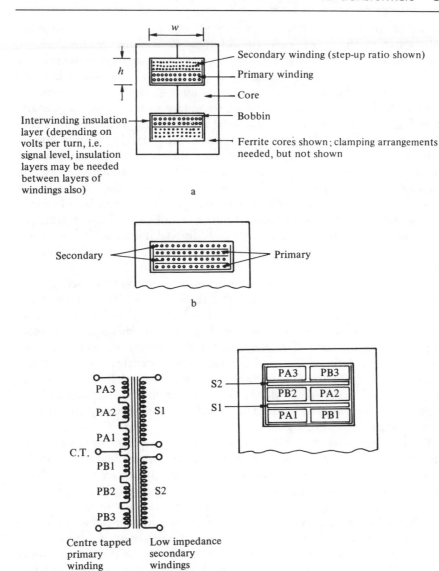

Figure 3.2 *Various winding arrangements.*
  a Basic transformer construction (bobbin and windings shown sectioned), showing winding width and height: w *and* h.
  b Scrap view showing windings sectionalized and interleaved for reduced leakage inductance.
  c Compound sectionalization.

However, given the minimum necessary number of turns, further steps are still possible. The most important of these is winding sectionalization. Figure 3.2a shows a transformer with the secondary overlaying the primary winding. Maximizing the winding width w and minimizing the height h will minimize the leakage inductance, but a much greater reduction can be achieved by interleaving sections of the secondary with those of the primary, as in Figure 3.2b, even if the additional interwinding insulation results in some increase in h. It is true that there will now be greater leakage inductance between one half of the primary and the other than previously, but this only results in a miniscule reduction in the total primary inductance: a similar comment applies to the secondary. However, the leakage inductance between the primary and secondary is greatly reduced, by a factor of n^2 where each winding is partitioned into n sections — at least in principle, ignoring the effect of increased h due to additional insulating layers. Where one of the windings is balanced, then side-to-side sectionalization can additionally be beneficially employed (Figure 3.2c).

At higher frequencies, e.g. RF, ferrite cores are universally used, as they maintain a high permeability at high frequencies while simultaneously exhibiting a low core loss. The high bulk resistivity of ferrite materials (typically a million times that of metallic magnetic materials, and often higher still in the case of nickel–zinc ferrites) results in very low eddy current losses, without the need for laminating. Ferrites for transformer applications are also designed to have very low coercivity, for low hysteresis loss: for this reason they are described as 'soft ferrites', to distinguish them from the high-coercivity hard ferrites used as permanent magnets in small loudspeakers and motors, etc.

For frequencies up to 1 MHz or so, MnZn (manganese zinc, sometimes known as 'A' type) ferrites with their high initial permeabilities (up to 10 000 or more) are usually the best choice. For much higher frequencies NiZn (nickel zinc or 'B' type) are often the best choice due to their lower losses at high frequencies, despite their lower initial permeability which ranges from 5 to 1000 or so for the various grades. At very high frequencies a further loss mechanism is associated with ferrite cores. Ferrite materials have a high relative permittivity, commonly as much as 100 000 in the case of MnZn ferrites. The electric field associated with the windings causes capacitive currents to circulate in the ferrite, which results in losses since the ferrite is not a perfect dielectric. The effect is less marked in NiZn ferrites — another reason for their superiority at very high frequencies.

For frequencies in the range 0.5 to 10 MHz, the preference for NiZn or MnZn ferrite is dependent on many factors, including the power level to be handled and the permissible levels of harmonic distortion and intermodulation. These and other factors are covered in detail in various sources, including References 1 and 2, whilst Reference 3 contains a wealth of information, both theoretical and practical. Table 3.1 gives typical values for

Table 3.1a   Managanese–Zinc ferrites for entertainment and industrial applications (Reproduced by courtesy of NEOSID Limited)

Parameter	Symbol	Standard conditions of test		Unit	F5	F6	F7	F8	F9	F11
Initial permeability (minimum)	μ_i	B→0 25°C	kHz	—	1600 1	1200 1	1800 100	1200 100	3500 1	500 100
Saturation flux density	B_{sat}	H = 796 A/m =10 Oe 25°C		mT	480	450	390	380	380	380
Loss factor (maximum)	$\dfrac{\tan \delta_{r\ +\ e}}{\mu_i}$	B→0 25°C	20 kHz 100 kHz 250 kHz 500 kHz 1 MHz	10^{-6}	— — — — —	— — — — —	8 — — — —	— — 30 110 —	— 15 — — —	— 20 — — 47
Temperature factor	$\dfrac{\Delta\mu}{\mu_i^2.\ \Delta T}$	B < 0.25 mT +25°C to + 55°C	10 kHz	$10^{-6}/°C$	—	—	0 to + 2	0 to +1	0 to +2	+0.5 to +2.5
Amplitude permeability (minimum)	μ_a	200 mT 25°C 400 mT 25°C 320 mT 100°C	50 Hz 50 Hz 50 Hz	—	3000 1000 1000	3000 1000 1000	— — —	— — —	— — —	— — —
Total power loss density (maximum)	—	200 mT 25°C 100°C	16 kHz 16 kHz	mW/cm³	110 120	150 150	— —	— —	— —	— —
Curie temperature (minimum)	θ_c	B < 0.25 mT	10 kHz	°C	200	180	150	130	135	220
Resistivity (typical)	ρ	1 V/cm 25°C		ohm cm	100	100	100	100	50	500

Table 3.1b   Nickel–zinc ferrites for entertainment and industrial applications (Reproduced by courtesy of NEOSID Limited)

Parameter	Symbol	Standard conditions of test		Unit	F13	F14	F16	F25*	F22	F29*
Initial permeability (±20%)	μ_i	B→0 25°C	1 MHz	—	650	220	125	50	19	12
Saturation flux density	B_{sat}	H = 796 A/m =10 Oe 25°C		mT	320	350	340	—	—	—
Loss factor (maximum)	$\dfrac{\tan \delta_{r\ +\ e}}{\mu_i}$	B→0 25°C	250 kHz 500 kHz 1 MHz 2 MHz 3 MHz 5 MHz 10 MHz 15 MHz 20 MHz 40 MHz 100 MHz 200 MHz	10^{-6}	50 65 130 — — — — — — — — —	— 40 42 50 — — — — — — — —	— — 60 — — 65 100 — — — — —	— — 50 50 55 65 75 100 125 300 — —	— — — — — — 300 — 330 500 — —	— — — — — — 100 — — — 200 1000
Temperature factor	$\dfrac{\Delta\mu}{\mu_i^2.\ \Delta T}$	B < 0.25 mT +25°C to + 55°C	10 kHz	$10^{-6}/°C$	1.5 (av'ge)	12—30	20—50	10—15	12—17	50 (av'ge)
Curie temperature (minimum)	θ_c	B < 0.25 mT	10 kHz	°C	180	270	270	450	500	500
Resistivity (typical)	ρ	1 V/cm 25°C		ohm cm	3.10^4	10^5	10^5	10^5	10^5	10^5

*These are perminvar ferrites and undergo irreversible changes of characteristics (μ increases and loss factors become much greater, especially at higher frequencies) if subjected to strong magnetic fields or mechanical shocks.

some of the more important parameters of typical MnZn and NiZn ferrites produced by one particular manufacturer, together with typical applications. The greater suitability of NiZn ferrites for higher frequencies is clearly illustrated. There are numerous manufacturers of ferrites and a selection of these (not claimed to be exhaustive) is given in Appendix 7.

The selection of a suitable low loss core material is an essential prerequisite to any successful wideband transformer design, but at least as much attention must be paid to the design of the windings. For wideband RF transformers, copper tape is often the best choice, at least for low impedance windings such as 50 Ω or less. This must be interleaved with insulating material, such as a strip of photographic mounting tissue (which, being waxed, sticks to itself when heated with the tip of an under-run soldering iron), or, for high power transformers, a high dielectric strength electrical tape such as PTFE. For a high impedance winding, such as the secondary of a 50 Ω to 600 Ω balun (balanced to unbalanced transformer), wire is the best choice. It can be enamelled, or in the case of a high-power transformer, PTFE insulated. A single layer is always preferable, if at all possible, as stacked layers exhibit a much inferior Q factor — resulting in increased insertion loss — and an embarrasing amount of winding self-capacitance, leading to problems at the top end of the band especially in high power transformers. A single-layer secondary winding in a balun is inherently symmetrical of itself, but the balance can be easily upset by electrostatic coupling from the signal in the primary winding, the 'hot' end of which will be in phase with one end of the balanced secondary winding and in antiphase with the other. However, the use of an interwinding screen results in an undesirable increase in spacing between the primary and secondary, resulting in increased leakage inductance. Where a full width copper tape primary underneath a solenoidal wirewound secondary is used, the solution is to use the earthy end of the primary itself as the screen, by making the start of the primary the 'hot' end, carrying the earthy end on beyond the lead-out for an extra half turn for symmetry.

Whether in the development or production phase, the degree of balance of a balun transformer needs to be checked to ensure all is well. Balance is measured in decibels and is defined as in Figure 3.3a, with a numerical example in Figure 3.3b: this is analysed into pure balanced and unbalanced components in Figure 3.3c. It can be seen that balance is defined independently of the transformer ratio. The balanced winding (usually regarded as the secondary) is shown in Figure 3.3 as having a centre tap connected to ground. Where neither the centre tap (if provided) nor any other part of the winding is connected to ground, the winding is said to be floating. In use, the balance achieved under these conditions is strongly influenced by the degree of balance of the load to which the transformer is connected. The balance of the transformer can conveniently be measured with the aid of a suitable balance pad. The purpose of such a pad is two-fold; firstly to terminate the secondary in its design impedance (e.g. 600 Ω), and secondly to provide a matched source, usually 50 Ω, for the measuring system. The major

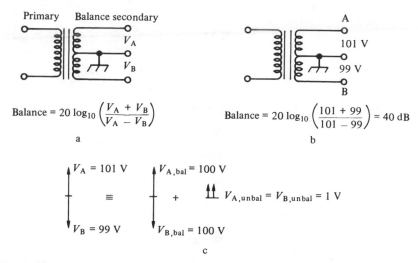

Balance secondary

$$\text{Balance} = 20 \log_{10} \left(\frac{V_A + V_B}{V_A - V_B} \right)$$

a

$$\text{Balance} = 20 \log_{10} \left(\frac{101 + 99}{101 - 99} \right) = 40 \text{ dB}$$

b

c

Figure 3.3 *Balanced transformer operation.*
a Definition.
b Example.
c Common mode components.

cause of any difference between the two half secondary voltages, particularly at the lower end of the balun's frequency range, is a difference in flux linkage with the primary. Because the difference is small compared with the total flux, the unbalanced component may be considered as arising from a negligibly small source impedance. The balance pad is used to pad this up to the characteristic impedance of the measurement system. Figure 3.4a shows the measurement set-up and Figure 3.4b shows balance pads for a number of common combinations of primary and secondary impedances. The insertion loss measured via the balun as in Figure 3.4a, less the allowance given in

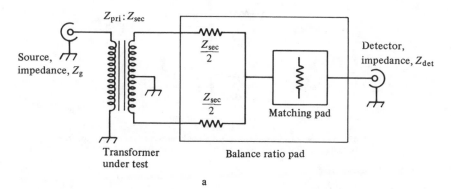

a

Figure 3.4 *Special pads for measuring balance ratios.*
a Balance measurement. Usually $Z_g = Z_{det} = 50 \; \Omega$ or $75 \; \Omega$.

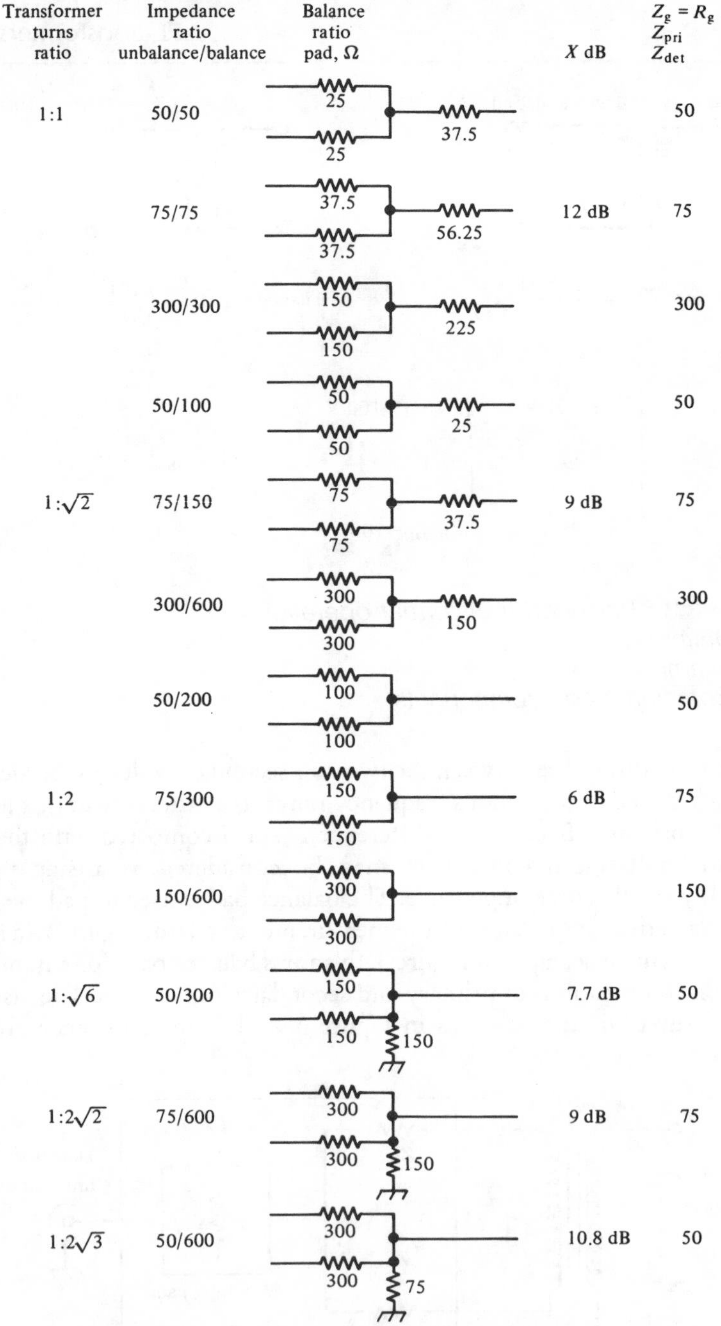

Transformer turns ratio	Impedance ratio unbalance/balance	Balance ratio pad, Ω	X dB	$Z_g = R_g$ Z_{pri} Z_{det}
1:1	50/50	25 / 25 / 37.5		50
	75/75	37.5 / 37.5 / 56.25	12 dB	75
	300/300	150 / 150 / 225		300
	50/100	50 / 50 / 25		50
1:$\sqrt{2}$	75/150	75 / 75 / 37.5	9 dB	75
	300/600	300 / 300 / 150		300
	50/200	100 / 100		50
1:2	75/300	150 / 150	6 dB	75
	150/600	300 / 300		150
1:$\sqrt{6}$	50/300	150 / 150 / 150	7.7 dB	50
1:2$\sqrt{2}$	75/600	300 / 300 / 150	9 dB	75
1:2$\sqrt{3}$	50/600	300 / 300 / 75	10.8 dB	50

X dB is the figure to be subtracted from the Insertion Loss of the transformer plus its Balance Ratio Pad to obtain the transformer balance.

b

Figure 3.4 *Continued*
b Balance pads for transformers of various ratios.

Figure 3.4b for the particular balance pad in use, gives the transformer balance ratio in decibels.

Returning for a moment to Figure 3.2a, it represents a sectional view of a transformer using ferrite 'E' cores. If the two ends of the primary winding are brought out on the same side of the core, then the primary will consist of a whole number of turns around the centre limb, and similarly for the secondary, which is normal good practice. Figure 3.2a could equally represent a ferrite transformer pot core with either two or four slots, and here again good practice is to use an integral number of turns, i.e. the two ends of any given winding (and any intermediate taps thereon) are brought out through the same slot, although of course different slots may be used for different windings. The core is dimensioned by the manufacturer to give equal flux density in the centre limb and each of the outer limbs when the windings consist of an integral number of turns. A half turn violates this condition, since the associated flux path is down one outer limb, returning through the centre and the other outer limb in parallel. In a high power transformer with only a few turns, the unequal flux density would reduce the power rating the transformer can handle if saturation in one of the limbs is to be avoided. Although we are concerned here only with transformers, it is worth pointing out that a half turn is even more undesirable in an inductor pot core, with its gapped centre limb. For every whole turn, the associated flux must pass through the centre limb with its air-gap, returning through the two or four outer limbs in parallel. With a half (or quarter or three-quarter) turn, the flux can pass down one or more outer limbs and back through other outer limbs, all ungapped. Thus a half turn may have substantially higher inductance than a whole turn, together with higher losses and a terrible temperature coefficient of inductance!

It was mentioned earlier that the useful LF (low frequency) response is set by the shunting effect of L_m across the transformed load resistance R', resulting in a -3 dB point (see Figure 3.5a) at that frequency where the reactance of L_m has fallen to half the characteristic impedance of the primary circuit. This is clear from Figure 3.5b where the matched source is shown in the alternative ideal current generator form, with everything normalized to unity. The LF response can be maintained down to a slightly lower frequency by connecting a suitable capacitor in series with the primary winding, as in Figure 3.5c. This can reduce the loss from 3 dB without the capacitor, to 2.5 dB with it — not a spectacular improvement but maybe enough to enable you to meet the specification requirement even though you cannot find a better core or squeeze another turn on. The problem is that the parallel combination of R' and L_m is equivalent (at any frequency) to the series combination of a resistor R'', less than R', and an inductance L'_m, less than L_m. The capacitor can only improve things marginally by tuning out L'_m; it cannot transform R'' back to R'. R' is of course equal to the characteristic impedance of the source and is thus the only value of load that

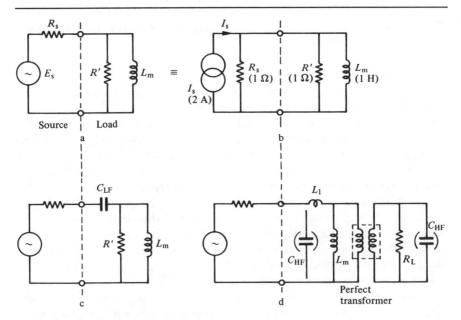

Figure 3.5 *Transformer bandwidth extension.*
  a Illustrating LF 3 dB point.
  b Shunt equivalent of a_s normalized to 1Ω.
  c Series C for LF extension.
  d Shunt C alternatives for HF extension.

can draw maximum power from the source. One could however choose R' to be deliberately mismatched to the source at mid band. The lower -3 dB point can then be extended down considerably by arranging that R'' is equal to the source resistance. This results in a second order Chebychev high-pass response, the degree of LF extension possible being set by the acceptable pass-band ripple. In a small-signal transformer, where bandwidth may be more important than efficiency, this scheme may well be worthwhile. Note that when a capacitor is used in series with the primary, the impedance presented to the source way below the band of interest rises towards infinity rather than falling towards a short circuit. This characteristic can be useful in some applications.

A similar marginal improvement can be had at the HF end of the transformer's range, where the response has fallen by 3 dB due to the increasing reactance of the leakage inductance. Here again capacitance can be used, this time in parallel with the transformer, to tune out the leakage inductance. Again, the 3 dB point can be improved to 2.5 dB, pushing up the -3 dB frequency by a small amount, or by rather more if a second order Chebychev low-pass response is acceptable. The capacitance can be connected either up- or down-stream of the leakage inductance, i.e. across the primary or secondary winding. In the latter case, it may well be possible to

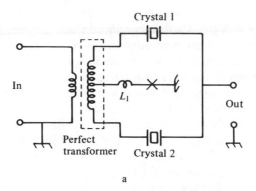

a

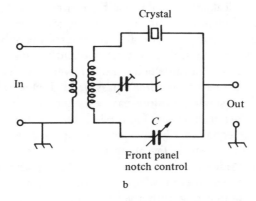

b

Figure 3.6 *This application requires the secondary voltages to be perfectly balanced.*
  a Half lattice crystal filter.
  b Economy version of a.

build the capacitance into the transformer, by using wire with thin insulation for the secondary, or possibly by using a multilayer winding.

There is one case where tuning can be used to overcome the deleterious effect of leakage inductance completely, admittedly only at one frequency — although that is no problem in this particular application. The application in question is a crystal filter. These are available very cheaply in standard frequencies such as 10.7 MHz, 21.4 MHz, 45 MHz, etc., being usually implemented with monolithic dual resonators, or even in the latest designs, quad resonators. However, this technology is not appropriate to small quantities of filters of a non-standard frequency. Here, a filter is more likely to use discrete crystals, the classical configuration being the lattice filter, using four crystals per section. The arrangement of Figure 3.6a is more economical, using only two crystals per section, with the aid of a balun transformer. In this instance it is essential that the centre tap of the balanced secondary winding be effectively earthed and that the voltages applied to the two crystals are exactly equal in amplitude and in antiphase. This not-withstanding the wildly unequal impedances of the two crystals across the band, bearing in mind that for optimum band-pass response, the two crystals have different series resonant frequencies. In this application, the problem is not the leakage inductance between primary and secondary, but that between the two halves of the secondary. The equivalent circuit can be drawn as two perfectly coupled half windings, with the leakage inductance in series with the centre tap lead-out. If the load impedances connected to the ends of the secondary, although varying with frequency, were always identical at any given frequency, the leakage inductance would be immaterial since no current would flow through it. Unfortunately this is not the case, but by inserting capacitance at point X in Figure 3.6a and tuning it to series resonance with the leakage inductance at the centre of the filter's pass band, the (inaccessible) junction of the perfectly coupled pair of windings is effectively shorted directly to earth. This short circuit is only effective at the resonant frequency of the leakage inductance and the inserted capacitance, but due to the L/C ratio of these being much lower than that of the crystals, it holds over the whole of the filter's pass band. Incidentally, if a simpler second-order filter (single pole low-pass equivalent) will suffice, the even more economical arrangement of Figure 3.6b may be used. Here, with the capacitance C set equal to C_0, the parallel capacitance of the crystal, a symmetrical response results. Tweaking C up or down in value will give a deep notch on one side of the response or the other, an arrangement popular at one time in amateur receivers, to notch out a strong CW signal when 'DXing', i.e. communicating with a very distant station.

Finally, no discussion of RF transformers would be complete without covering line transformers. These were popularized by a paper published as long ago as 1959 [4], although the idea was not new even then, Ruthroff's paper containing five references to earlier work. The basic principle of the transmission line transformer is to cope with the leakage inductance and winding capacitance by making them the distributed L and C of an RF line; a neat idea, although in the process dc isolation between primary and secondary is lost, in many cases. Figure 3.7a shows a 1:1 inverting transformer: the impedance of the line should equal the nominal primary and secondary impedance. If this is 50 Ω, then miniature coax can con-veniently be used. Wire 1–2, the inner, carries (in addition to the load current drawn by R, which returns through 4–3 and hence produces no net flux on the core) the magnetizing current needed to establish the flux on the core. This magnetizing current returns via the connection between the earthy end of the load and the earthy end of the source. The flux induces in series with both outer and inner a voltage equal to the voltage applied between points 1 and 3 (ground). The arrangement can be regarded as an ideal inverting transformer in series with a length of transmission line. The higher the permeability of the core, the fewer turns will be needed to obtain

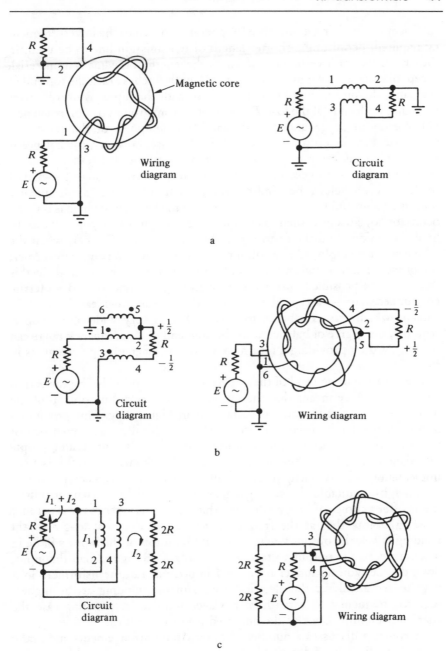

Figure 3.7 *Various examples of line transformers.*
a Reversing transformer.
b Unbalanced to balanced transformer.
c 4:1 Impedance transformer.

sufficient magnetizing inductance for operation down to the lowest frequency required, permitting a shorter length of transmission line to be used. In the case of the 1:1 inverting transformer, the length of the line is immaterial, except of course insofar as if the electrical length reaches $\lambda/2$ at the top end of the band, the output will be back in phase with the input. Ruthroff states that since both ends of the load R are isolated from ground by coil reactance, either end can be grounded, and that if the midpoint of the resistor is grounded then the output is balanced. In this case, however, the balance is not complete, as some magnetizing current is still needed (exactly half as much as in the inverting case), and this must now return through one-half of the load. Nevertheless, the winding arrangement of Figure 3.7a is frequently used as a balun and proves satisfactory where the frequency range is only an octave or so, since it is then easy to provide enough primary inductance to hold the residual unbalance to acceptable proportions. Further, when the arrangement is employed as a balun rather than as an inverting transformer, the phase relation between input and output is usually immaterial. In this case it may be possible to use a long enough length of line to render a ferrite core unnecessary — a typical example is the coaxial downlead from a TV antenna which acts as a balun for free. Where a very wideband balun is required, the degree of balance at the bottom end of the frequency range can be preserved by providing a return route for the magnetizing current, as in Figure 3.7b.

The isolation of one end of the line from the other provided by the end to end coil reactance means that the output can be stacked up on top of the input, to give twice the output voltage, as in Figure 3.7c. This provides a non-inverting 4:1 impedance ratio transformer. Ideally, the impedance of the line used should be the geometric mean of the input and output impedances, i.e. 100 Ω in the case of a 50 Ω to 200 Ω transformer: this is easily implemented with two lengths of self-fluxing enamelled magnet wire twisted together, by a suitable choice of gauge, insulation thickness (wire manufacturers offer a choice of fine, medium or thick) and a number of turns per inch twist [5]. Note that at the frequency where the electrical length of the winding is $\lambda/4$, the output voltage stacked up on top of the input will be in quadrature, so the output voltage will be only 3 dB higher than the input, not 6 dB, i.e. you no longer have a 4:1 impedance ratio transformer. So it pays to try and keep the electrical length of the winding at the highest required frequency to a tenth of a wavelength or less; in this case the characteristic impedance of the line used is not too critical.

Reference 4 discusses a number of other circuit arrangements and many others have since been described, mostly limited to certain fixed impedance ratios such as 4:1, 9:1, and 16:1, sometimes combined with an unbalanced to balanced transition or vice versa. Reference 6 is useful, while Reference 7 discusses slipping an extra turn or two onto the core, to obtain ratios intermediate between those metnioned above. Line transformers can usefully provide bandwidths of up to 10 000:1, given a suitable choice of core. However, where a much more modest bandwidth is adequate, it may be possible to omit the core entirely, e.g. the case of a TV downlead acting as a balun, as already mentioned. Freed from the constraints of a core, it is possible to consider using a non-constant impedance line. In particular, balanced transmission lines having a characteristic impedance increasing exponentially with distance were described in patents lodged in America, Germany and Australia in the 1920s. Reference 8 describes a quasi-exponentially tapered line transformer providing a 200 Ω to 600 Ω transition over the range 4 to 27.5 MHz. True, it is 41 m long, but then it does consist of nothing but wire (plus a few insulating supports) and has a rating of 20 kW continuous, 30 kW peak.

References

1. Snelling, E. C. *Soft Ferrites, Properties and Applications*, Butterworths, London (1969)
2. Snelling, E. C. and Giles, A. D. *Ferrites for Inductors and Transformers*, Research Studies Press Ltd, UK, John Wiley and Sons, USA (1983)
3. DeMaw, M. F. *Ferromagnetic-Core Design and Application Handbook*, Prentice Hall, USA (1981)
4. Ruthroff, C. L. *Some Broad-Band Transformers*, Proceedings of the I.R.E., pp. 1337–42 (August 1959)
5. Lefferson, P. Twisted magnet wire transmission line. *IEEE Transactions on Parts, Hybrids and Packaging*, **PHP-7**(4), pp. 148–54 (December 1971)
6. Granberg, H. *Broadband Transformers and Power Combining Techniques for RF*, Motorola Application Note AN-749 (1975)
7. Krauss, H. L. and Allen, C. W. Designing toroidal transformers to optimize wideband performance. *Electronics*, 16 August 1973
8. Young, S. G. H.F. exponential-line transformers. *Electronic and Radio Engineer*, 40–44 (February 1959)

4
Couplers, hybrids and directional couplers

This chapter describes some further important passive components. Hybrids are based upon transformer action, whilst directional couplers depend upon capacitive coupling in addition. First a look at simple resistive couplers or 'splitters'. These can be used to split a signal between two outputs, in any desired ratio. Figure 4.1a shows three-way resistive splitters which provide a 50 Ω match at each port, provided that the other ports are correctly matched. Any port can be used as the input and the outputs at the other two are each 6 dB down on the input and both are in phase with it. There is thus 3 dB more attenuation at each output than with an ideal hybrid divider, which has no internal losses. There is also only 6 dB of isolation between the two output ports, but against these disadvantages resistive splitters/combiners are cheap and operate from dc to microwave frequencies. If additional loss from input to output can be accepted, the isolation between outputs increases faster than the through loss. Thus in Figure 4.1b, the loss from port A to B (or C) is 20 dB, but the isolation between ports B and C is 34 dB. Other designs (such as 10 dB through with 14 dB isolation) are simply designed by adding T pads to ports B and C of the basic 6 dB splitter of Figure 4.1a (4 dB in this case), and then combining the series resistors. The pad of Figure 4.1b is useful for combining two signals without them intermodulating, by maintaining high isolation between them, e.g. audio tones for two-tone transmitter testing, or two RF signals for intermodulation tests. Symmetrical pads with any number of ways are easily designed. Figure 4.1c shows a six-port 50 Ω splitter, the loss from an input to any output being 14 dB. Such multiport couplers are useful for hardwired signal-path testing of a radiocommunications net with N transceivers. Where two unequal outputs are required, the through loss to the greater of the two outputs can be less than 6 dB. Figure 4.1d shows a resistive divider for use as a 'signal sniffer', e.g. to sample the output of a transmitter for application to a spectrum analyser. The output at port C is 40 dB down on that at port B. The loss from port A to B is less than 0.2 dB. In practice, the two 0.5 Ω

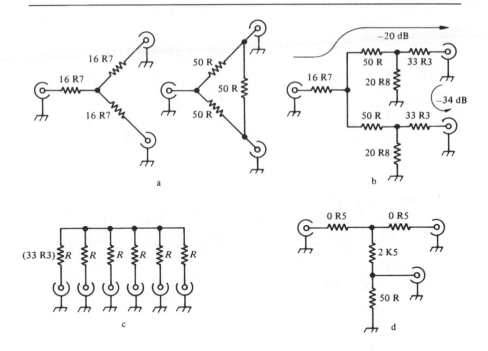

Figure 4.1 *Resistive couplers (50 Ω system).*
 a 6 dB Symmetrical two-way (three port) splitters/combiners.
 b 20 dB Half-symmetrical splitter/combiner.
 c Five output splitter (N = 6) for any N: R = 50 – 100/N (for 50 Ω system), loss = 20 log$_{10}$ (N – 1). For N = 6, loss = isolation = 14 dB.
 d 40 dB Signal sniffer (see text).

resistors would probably be omitted. The design of assymetric dividers for splitting losses which differ by only a few decibels is tedious; if 3 dB attenuation is acceptable in the main path then it is simpler to add a pad giving the required difference in attenuation to the output of a Figure 4.1a type splitter.

A hybrid can divide the input signal power between two outputs with negligible loss, each output being 3 dB down on the input. The basic hybrid circuit is shown in Figure 4.2a. If a signal is applied at port A, it will be divided equally between ports B and C whilst no power is delivered to port D (which could therefore be loaded with any termination from a short to an open circuit) as can be seen from the symmetry of the circuit, given that ports B and C are both terminated in 50 Ω. The outputs at ports B and C are in antiphase and the arrangement is known as a 180° hybrid (port D is often terminated internally in 25 Ω and only ports A, B and C made available to the user). The corollary is that if two identical signals of equal amplitude but

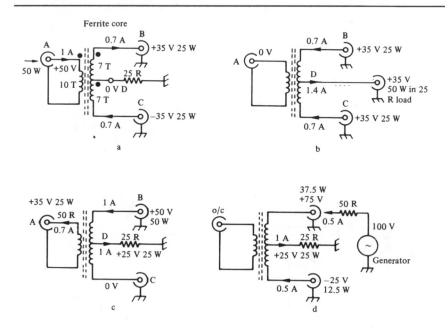

Figure 4.2 *The basic hybrid coupler.*
  a 180° hybrid, driven from 50 Ω matched source, P_{in} = 50 W.
  b In-phase power combining (see text)
  c Signal applied to port B.
  d As c, but port A open circuit. Matched source sees a load with 3:1 VSWR.

180° out of phase are applied to ports B and C, all of the available power is combined and delivered to port A, port D again being isolated. If, however, the two identical signals were in phase (Figure 4.2b), the currents in the centre tapped winding would produce no net flux on the core, so that port A is isolated and all the power is delivered to port D. If this is terminated with a 25 Ω load, then since ports B and C each supply half of the power, each will 'see' a 50 Ω termination. The corollary is that if a signal is applied at port D, it will be divided equally between ports B and C, the outputs being in phase, with port A isolated. This arrangement is known as a 0° hybrid: port A may be terminated internally in 50 Ω and an autotransformer is usually fitted to transform port D to 50 Ω. The 180° hybrid is cheaper as an autotransformer is not needed. Sometimes all four ports are brought out, giving a 'sum and difference hybrid'.

Figure 4.2c shows what happens if a signal is applied to port B. The input power divides equally between port A and 'port D' — a 25 Ω resistor in the case of a 180° hybrid — with port C isolated. The split between ports A and

D is almost perfect, the small difference component of current required to supply the magnetizing flux on the core being in quadrature. Thus for a correctly terminated four port hybrid, the power always splits equally between ports adjacent to the input port, the opposite port being isolated. Figure 4.2d shows what happens if one of the adjacent ports is mismatched — here port A is open circuit. A current of 0.5 A flows into port B and the currents at ports C and D can only be as shown, since there must be ampere–turn balance in the centre-tapped winding. So the output voltages and powers at ports B and D can be marked in. A total of 37.5 W is supplied to port B, and the voltage there is 75 V: the source sees a load of 150 Ω instead of the designed load of 50 Ω. Note that even for this extreme mismatch of one adjacent port, the power in the other is totally unaffected, and still twice that in the 'isolated' port: if the mismatch at adjacent ports is small, a hybrid provides high isolation at the fourth port. Most importantly, the fact is that open-circuiting (or short-circuiting) port A has no effect whatever on the power delivered by port B to port D, indicating perfect mutual isolation between the two opposite ports adjacent to the input port (if and only if the source impedance is an ideal 50 Ω).

A five-port hybrid divides power equally between four output ports, maintaining high isolation between them. It consists of two hybrids connected to opposite outputs of a third Figure 4.2a type hybrid and can equally well be used to combine the power outputs of, say, four amplifier stages in a solid state transmitter. Usually the difference ports of the three constituent hybrids are terminated internally. Further levels of build-up can provide 8- or 16-way couplers, etc. Occasionally the number of ways required is not a power of two. Figure 4.3 shows a hybrid which splits the input power three ways. It is instructive to work out what happens if one of the output ports is mismatched, port C open circuit for example. Remember that as the primaries of the three transformers are in series, the secondary currents cannot differ substantially, but that as the magnetizing current is small (and in quadrature), the primary voltages can differ. It turns out that on open circuiting port C, the outputs at ports B and D are unchanged, but that in each case one-third of the current is provided via one of the 150 Ω resistors from the centre transformer. Furthermore the load seen by the generator rises to 2Z, the power supplied by it falls by 1/9th and the voltage at the input port rises by a third. Full marks if your analysis comes up with these results: hint, the secondary voltage of the centre transformer doubles. Figures 4.2 and 4.3 together enable low-loss high-isolation splitting or combining arrangements for 2, 3, 4, 6, 8 or 9 outputs. A five-way split can be achieved rather like Figure 4.3 but using five transformers with primaries in series: a terminating resistor is required between each possible pair of secondary outputs. The arrangement is unwieldy and even more so for seven or more ways. So for a seven-way split, it is usually better to use an eight-output hybrid and simply terminate off the unused output. For

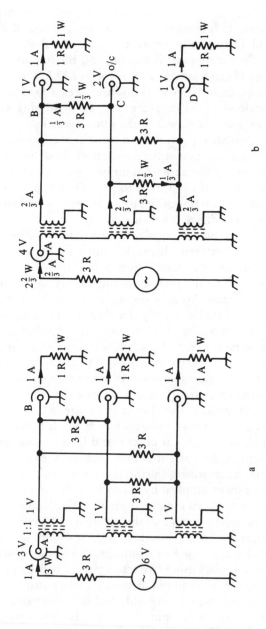

Figure 4.3 *Three-output hybrid. (Normalized to 3 Ω in, 1 Ω out to illustrate operation. For a 50 Ω hybrid at all ports, transformer ratios are each 4:7.)*
a Normal operation.
b One output open circuit, other outputs unaffected — ideally infinite isolation between output ports.

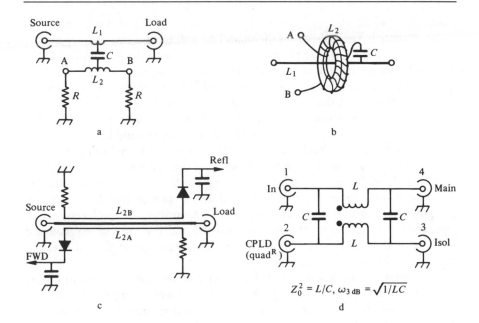

Figure 4.4 *Couplers.*
a–c Directional.
d Quadrature (see text).

combining, e.g. of transmitter modules, it is better to design around a power of two (and/or three) modules from the outset.

The coupler of Figure 4.1d could be used to obtain a low level sample of a high power signal, e.g. for measurement purposes. The same output at port C results whether the power in the 'main line' flows from port A to B or vice versa. In a directional coupler the transfer of power from one port to another is dependent upon the direction of power in the main line. The operation of one type is as follows (see Figure 4.4a). Power from a source, e.g. a transmitter, flows through the primary of a current transformer L_1, e.g. to a (hopefully) matched antenna presenting a 50 Ω load. It is important to note that the reactance of L_1 is very low compared to 50 Ω, so that the current flowing is determined solely by the power available from the source and the impedance of the load. Imagine for the moment a 50 Ω source and that the load is a short circuit: then the current flowing will induce a quadrature voltage in L_2 proportional to the rate of change of the current. Half of the voltage will appear at A and the other half at B, in antiphase, since the two earthed resistors R are equal and form a balanced bridge. The capacitor C will have no effect, as there is no voltage at the centre tap of L_2, nor at L_1 due to the shorted load. Now imagine the load is open circuit: no current flows through L_1 so no voltage is induced in L_2, so points A and B must be at the same potential. The voltage on the main line will force a leading (capacitive) current through C, whose reactance is much higher than R. Suppose C has been selected so that the voltage produced at A is the same as when the load was short circuited. Now, when a matched load is connected, the components of voltage at A due to inductive and capacitive coupling will add, while those at B will cancel out. If the direction of flow of power in the main line were reversed, the voltages at B would add and there would be no voltage at A. With any value of load, the voltage at A is proportional to the forward power and that at B to the reverse power, so if diode detectors are connected at A and B, we have a means of monitoring the forward power supplied by the source and reverse power reflected by a mismatched load, e.g. for purposes of measurement and control in a transmitter. As the frequency of operation is raised, both the current-induced and the capacitively-coupled voltages will rise pro rata. Consequently the detected voltages will rise, but the directivity is maintained.

The construction of a directional coupler can take many forms: in Figure 4.4b a toroidal core surrounding the main line (a single turn primary) is used. In Figure 4.4c separate lines L_{2A} and L_{2B} are used as secondaries to monitor forward and reverse power separately. The dimensions and spacings of the three lines are chosen to give the appropriate ratio of capacitive to inductive coupling. It is important that the coupled lines are short compared to a wavelength, so that the capacitive coupling can be considered as a lumped component. This results in the signal coupled into the measuring circuit being only a tiny fraction of the through energy, a limitation which is quite acceptable, indeed desirable, in this application. When two lines are close spaced over an appreciable fraction of a wavelength, much tighter coupling can be achieved. If the lines are one-quarter of a wavelength long at the operating frequency, a 3 dB split of power between the main and coupled lines can be achieved, the main and coupled outputs being in quadrature. This technique is conveniently implemented at UHF using 'microstrip' or 'stripline' lines. A microstrip line consists of a track on a printed circuit board (the other side of which is covered in copper ground plane), the width required to give a 50 Ω impedance depending upon the thickness and dielectric constant of the PCB material [1, 2]. Stripline is similar but covered with a second PCB carrying just a copper ground plane. Using this technique, quadrature couplers operating at frequencies as low as VHF are available, the coupled lines being 'meandered' on the surface of the PCB, for compactness. Bandwidth is typically 10% for ±0.6 dB variation in amplitude between the main and quadrature outputs. More complicated structures offer quadrature couplers with $1\frac{1}{2}$ octave bandwidth [3] whilst quadrature couplers covering 2–32 MHz have been designed by Merrimac. At these frequencies, quadrature couplers use lumped components, the basic narrowband section being as in Figure 4.4d. The two inductors L are wound using bifilar wire to give 100% coupling, and Figure 4.4d gives the component values in terms of the design impedance level and centre frequency.

References

1. Tam, A, *Principles of Microstrip Design*, RF Design, pp. 29–34 (June 1988) (With further useful references)
2. *Microwave Filters, Impedance Matching Networks and Coupling Structures*, Matthei, Young and Jones, McGraw-Hill, 1964
3. Ho, C. Y. *Design of Wideband Quadrature Couplers for UHF/VHF*, RF Design, pp. 58–61 (November 1989) (With further useful references)

5

Active components for RF uses

The simplest semiconductor active device for RF applications is the diode, which like its thermionic forebear conducts current in one direction only. Arguably, semiconductor diodes are not active devices, simply non-linear passive ones, but their mode of operation is so closely linked with that of the transistor that they are usually considered together. The earliest semiconductor diode was of the point contact variety — the user-adjusted crystal and cat's whisker used in the early days of wireless. Later, new techniques and materials were developed, enabling robust pre-adjusted point contact diodes useful at radar frequencies to be produced. Germanium point contact diodes are still produced and are useful where a diode with low forward voltage drop at currents of a milliampere or so, combined with low reverse capacitance, is required. However, for the last 25 years, silicon has been the preferred material for semiconductor manufacture for both diodes and transistors, whilst point contact construction gave way to junction technology even earlier. Figure 5.1a shows the I/V characteristics of practical diodes. Silicon is one of the substances which exists in a crystalline form with a cubic lattice. When purified and grown from the melt as a single crystal, it is called intrinsic silicon and is a poor conductor of electricity, at least at room temperature. However, if a few of the silicon atoms in the atomic lattice are replaced by atoms of a pentavalent substance such as phosphorus (which has five valence electrons in its outer shell, unlike the four electrons of quadravalent silicon), then there are spare electrons with no corresponding electron in an adjacent atom with which to form a bond pair. These spare electrons can move around in the semiconductor lattice, rather like the electrons in a metallic semiconductor, though the conductivity of the material is lower than that of a metal, where every single atom provides a free electron. The higher the 'doping level', the more free electrons and the higher the conductivity of the material, which is described as N type, indicating that the flow of current is due to negative carriers, i.e. electrons. P type silicon is obtained by doping the monocrystalline silicon lattice with a

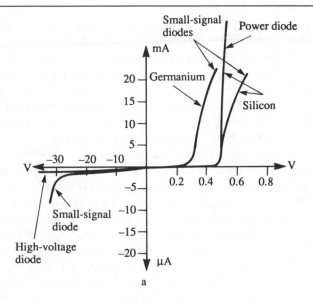

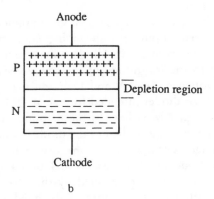

Figure 5.1 *Semiconductor diodes.*
a I/V characteristics.
b Diagrammatic representation of PN diode, showing majority carriers and depletion region.

sprinkling of trivalent atoms such as boron. Where one of these exists in the lattice next to a silicon atom, the latter has one of its four outer valence electrons 'unpaired' — a state of affairs described as a hole. If this hole is filled by an electron from a silicon atom to the right, then whilst the electron has moved to the left, the hole has effectively moved to the right. It turns out that spare electrons in N type silicon are more mobile than holes in P type, which explains why very high frequency transistors are more easily made as NPN types.

Figure 5.1b shows diagrammatically the construction of a silicon diode, indicating the lack of carriers (called a depletion layer) in the immediate vicinity of the junction. Here, the electrons from the N region have been attracted across to fill holes in the P region. This disturbance of the uniform charge pattern that should exist throughout the N and P regions represents a potential barrier which prevents further electrons migrating across to the P region. When the diode is reverse biased, the depletion layer simply becomes more extensive. The associated redistribution of charge represents a transient charging current, so that a reverse biased diode is inherently capacitive. If a forward bias voltage large enough to overcome the potential barrier is applied to the junction, about 0.6 V in the case of silicon, then a forward current will flow. The incremental or slope resistance r_d of a forward biased diode at room temperature is given approximately by $25/I_a$ Ω, where the current through the diode I_a is in milliamperes. Hence the incremental resistance at 10 µA is 2K5, at 0.1 mA is 250 Ω and so on, but bottoming out in the case of a small-signal diode at a few ohms, where the bulk resistance of the semiconductor material and the resistance of leads, bond pads, etc., comes to predominate.

The varactor diode or varicap is a diode designed solely for reversed biased use. A special doping profile giving an abrupt or 'hyperabrupt' junction is used. This results in a diode whose reverse capacitance varies widely according to the magnitude of the reverse bias. The capacitance is specified at two voltages, e.g. 1 V and 15 V and may provide a capacitance ratio of 2:1 or 3:1 for diodes intended for use at UHF up to 30:1 for types intended for tuning in AM radios. In these applications, the peak-to-peak amplitude of the RF voltage applied to the diode is small compared with the reverse bias voltage, even at minimum bias where the capacitance is maximum. So the diode behaves like a normal mechanical variable capacitor, except that the capacitance is controlled by the reverse bias voltage rather than by a rotary shaft. Tuning varactors are designed to have a low series loss r_s, so that they exhibit a high quality factor Q over the recommended range of operating frequencies. Another use for varactors is as frequency multipliers. If an RF voltage with a peak-to-peak amplitude of several or many volts is applied to a reverse biased diode, its capacitance will vary in sympathy with the instantaneous RF voltage. Thus the device is behaving as a non-linear capacitor, and as a result the RF current through it will contain harmonic components which can be extracted by suitable filtering. A non-linear resistance would also generate harmonics, but the varactor has the advantage over a non-linear resistor of not dissipating any of the drive energy.

The P type/Intrinsic/N type or PIN diode is a PN junction diode, but fabricated with a third region of intrinsic (undoped) silicon between the P and N regions. When forward biased by a direct current it can pass RF signals without distortion, down to some minimum frequency set by the lifetime of the carriers, holes and electrons, in the intrinsic region. As the forward current is reduced, the resistance to the flow of the RF signal is increased, but it does not vary over a half cycle of the signal frequency. As the direct current is reduced to zero the resistance rises towards infinity: when the diode is reverse biased only a very small amount of RF current can flow, via the diode's reverse capacitance. The construction ensures that this is very small, so that the PIN diode can be used as an electronically controlled RF switch or relay. It can also be used as a variable resistor or attenuator, by adjusting the amount of forward bias current. An ordinary PN diode can also be used as an RF switch, but it is necessary to ensure that the peak RF current, when on, is smaller than the direct current, otherwise waveform distortion will occur. It is the long 'lifetime' (defined as the average length of time taken for holes and electrons in the intrinsic region to meet up and recombine, so cancelling each other out) which enables the PIN diode to operate as an adjustable linear resistor, even when the peaks of the RF current exceed the direct current.

When a PN diode which has been carrying direct current in the forward direction is suddenly reverse biased, the current does not cease instantaneously. The charge has first to redistribute itself to re-establish the depletion layer. Thus for a very brief period, the reverse current flow is much greater than the steady state reverse leakage current. The more rapidly the diode is reverse biased, the more rapidly the charge is extracted and the larger the transient reverse current. Snap-off diodes are designed so that the end of the reverse recovery pulse is very abrupt, rather than the tailing off observed in ordinary PN junction diodes. It is thus possible to produce very short sharp current pulses which can be used for a number of applications, such as high order harmonic generation (turning a VHF or UHF drive current into a microwave signal) or operating the sampling gate in a sampling oscilloscope.

Small-signal Schottky or 'hot carrier' diodes operate by a fundamentally different form of forward conduction. As a result of this, there is virtually no stored charge to be recovered when they are reverse biased, enabling them to operate efficiently as detectors or rectifiers at very high frequencies. Zener diodes conduct in the forward direction like any other diode, but they also conduct in the reverse direction and this is how they are usually used. At low reverse voltages a zener diode conducts only a small leakage current, like any other diode, but when the voltage reaches the nominal zener voltage the diode current increases rapidly, exhibiting a low incremental resistance. Diodes with a low breakdown voltage — up to about 4 V — operate in true zener breakdown: this conduction mechanism exhibits a small negative temperature coefficient ('tempco'). Higher voltage diodes rated at 6 V or more operate by a different mechanism, called avalanche breakdown, which has a small positive tempco. In diodes rated at about 5 V, both mechanisms occur, resulting in a very low or zero tempco. However, the lowest slope resistance is found in diodes rated at about 7 V. Zener diodes can be used to stabilize the dc operating conditions in an RF power amplifier. Zener diodes can also usefully be employed as RF noise sources and a very few are actually specified for this purpose. It is necessary to select a diode where the noise output level is reasonably independent of frequency over the desired operating range, and stable also with respect to operating current, temperature and life. Suitable diodes can provide a useful output (say 10 to 15 dB above thermal) up to 1 GHz.

Like diodes, bipolar transistors first appeared as point contact types, though all current production is of junction devices. However, the point contact structure is preserved to this day in the symbol for a transistor (Figure 5.2a). Figure 5.2b shows diagrammatically the structure of an NPN bipolar transistor: it has three separate regions. With the base (a term dating from point contact days) short circuited to the emitter, no current can flow in the collector, since the collector/base junction is a reverse biased diode, complete with depletion layer as shown. The higher the reverse voltage, the wider the depletion layer, which is found mainly on the collector side of the junction as the collector is more lightly doped than the base. In fact, the pentavalent atoms which make the collector N type are found also in the base region. The base is a layer which has been converted to P type by substituting so many trivalent (hole donating) atoms into the silicon lattice, e.g. by diffusion or ion bombardment, as to swamp the effect of the pentavalent atoms. So holes are the majority carriers in the base region, just as electrons are in the collector and emitter regions. The collector junction then turns out to be largely notional: it is simply that plane on the one side (base) of which holes predominate whilst on the other (collector) electrons predominate. Figure 5.2c shows what happens when the base emitter junction is forward biased. Electrons flow from the emitter into the base region and simultaneously holes flow from the base into the emitter. The latter play no useful part in transistor action: they contribute to the base current but not to the collector current. Their effect is minimized by doping the emitter a hundred times (or more) more heavily than the base, so that the vast majority of the carriers traversing the base/emitter junction consists of electrons flowing from the emitter into the base. Some of these electrons combine with holes in the base and some flow out of the base, forming the greater part of the base current. Most of them, being minority carriers (electrons in what should be a P type region) are swept across the collector junction by the electric field existing across the depletion layer. This is illustrated in diagrammatic form in Figure 5.2c, while Figure 5.2d shows the collector characteristics of a small-signal NPN transistor. It can be seen that for small values of base (and collector) current, the collector voltage has little effect upon the amount of current flowing, at least for collector/emitter

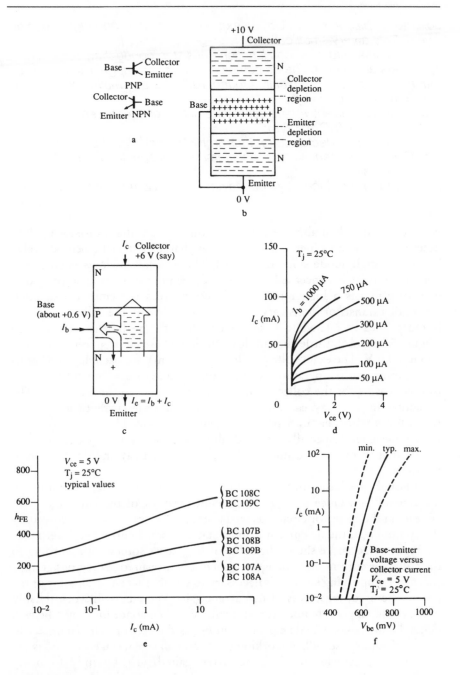

Figure 5.2 *The bipolar transistor.*
a Bipolar transistor symbols.
b NPN junction transistor, cut-off condition. Only majority carriers are shown. The emitter depletion region is very much narrower than the

collector depletion region because of no reverse bias and higher doping levels. Only a very small collector leakage current I_{cb} flows.

c NPN small-signal silicon junction transistor, conducting. Only minority carriers are shown. The dc common emitter current gain is $h_{FE} = I_c/I_b$, roughly constant and typically around 100. The ac small-signal current gain is $h_{ie} = dI_c/dI_b = i_c/i_b$.

d Collector current versus collector/emitter voltage, for an NPN small-signal transistor (BC 107/8/9).

e h_{FE} versus collector current for an NPN small-signal transistor.

f Collector current versus base/emitter voltage for an NPN small-signal transistor.

(Parts d to f reproduced by courtesy of Philips Components Ltd.)

voltages greater than about +1.5 V. For this reason, the transistor is often described as having a 'pentode like' output characteristic (the pentode valve has a very high anode slope resistance). This is a fair analogy as far as the collector circuit is concerned, but there the similarity ends. The pentode's control grid has a high input impedance whereas the emitter/base input circuit of a transistor looks very much like a diode, and the collector current is more linearly related to base current than to the base/emitter voltage (Figure 5.2e and f). Little current flows until the base/emitter voltage reaches about +0.6 V. The exact voltage falls by about 2 mV for each degree Celsius rise in transistor temperature, whether this be due to the ambient temperature increasing, or the collector dissipation warming up the transistor. The reduction in V_{be} may cause an increase in collector current, heating the transistor up further, in a potentially vicious circle. It thus behoves the circuit designer, especially when dealing with RF power transistors, to ensure that this process cannot lead to thermal runaway and destruction of the device.

Although the base/emitter junction behaves like a diode, exhibiting an incremental resistance of $25/I_e$ at the emitter, most of the emitter current appears in the collector circuit, as we have seen. The ratio I_c/I_b is denoted by the symbol h_{FE}, the dc current gain or static forward current transfer ratio. As Figure 5.2d and e show, the value of h_{FE} varies somewhat according to the collector current and voltage at which it is measured. When designing a transistor amplifying stage, it is necessary to ensure that any transistor of the type to be used, regardless of its current gain, V_{be}, etc., will work reliably over a wide range of temperatures: the no-signal dc conditions must be well defined and stable. The dc current gain h_{FE} is the appropriate parameter to use for this purpose. When working out the small signal stage gain, h_{fe} is the appropriate parameter; this is the ac current gain dI_c/dI_b. Usefully, for many modern small signal transistors there is little difference in the value of h_{FE} and h_{fe} over a considerable range of current, as can be seen from Figures 5.2e and 5.3a (allowing for the linear vertical axis in the one and logarithmic in the other).

The performance of transistors can be described by a number of ways, some implying a particular model of the transistor's internal circuit as in Figure 5.3b, while others simply relate conditions at the input port to those at the output. For use at the higher RF frequencies, certainly above 10 MHz say, the most useful approach is undoubtedly using *scattering parameters* (or *s*-parameters). These are so called as they involve measuring the voltage reflected or scattered at input or output port in a matched system, for a given incident voltage. They are dealt with in detail in Appendix 2. However, of the many other sets of parameters used to describe transistor function, historically one of the most important is the *hybrid parameter* set. This uses a simple model not presupposing an internal circuit of the transistor (see Figure 5.4a and b). h_{11} is the input impedance and h_{21} the forward current transfer ratio, both measured with the collector short-circuited at ac, while h_{22} is the output admittance and h_{12} the voltage feedback ratio (dv_1/dv_2), both measured with the input open circuit to ac. This set of parameters is known as the hybrid parameters (or *h*-parameters) due to the mixture of units, impedance, admittance and pure ratios. A transistor can be used as an amplifier in three fundamentally different circuit configurations, but there is one feature common to all of these. Having only three leads, one of the electrodes of a transistor amplifier must be common to both the input circuit and the output circuit, as indicated by the dotted line in Figure 5.4b. Figure 5.3c shows a common emitter small-signal amplifier using the BC109, a transistor designed originally as a low-noise AF amplifier, but useful in not too demanding RF circuits up to several tens of megahertz. When employed in the common emitter circuit, h_{21} is known as h_{fe}, which we have already met. Figure 5.4c shows h_{ie}, h_{re} and h_{oe}, the common emitter values of h_{11}, h_{12} and h_{22} respectively, for the BC109. These parameters are for operation at the standing values of collector current and voltage indicated, at 1 kHz. At this low frequency, there is negligible phase shift through the transistor under the prescribed measurement conditions, so the parameters are all real, not complex. Using these parameters, the low-frequency performance of a common emitter stage such as in Figure 5.3c can in principle be calculated exactly. However, the *h* parameters will vary with collector current and voltage (the graphs give data for only two spot values of collector emitter voltage) and in any case, are only typical values. In fact, for all the parameter sets mentioned in the text books, only a few are quoted in manufacturers' data, and maximum and minimum data are even scarcer. The advantage of *s* parameters is that they do not involve measurements made with a port terminated in open or short circuit, these being extremely difficult to implement precisely at RF. With *s* parameter measurements, the source and load impedance is 50 Ω, provided by the test ports of a network analyser.

The common emitter configuration of Figure 5.3c offers potentially the highest gain of the three configurations (the actual gain will depend more on the circuit than the transistor) because there is current gain and, if the collector circuit load impedance is higher than the stage's input impedance,

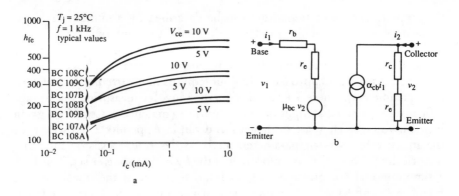

a

b

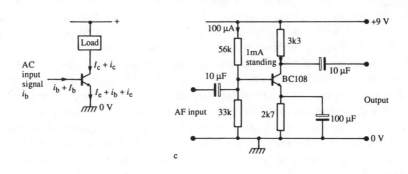

c

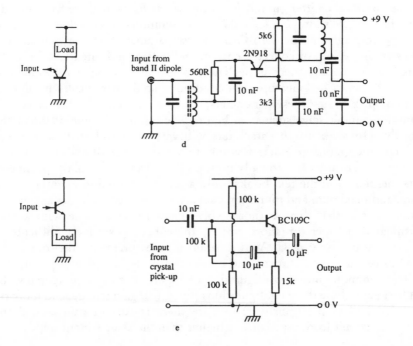

d

e

Figure 5.3 *Small-signal amplifiers.*
  a h_{fe} versus collector current for an NPN small-signal transistor of same type as in Figure 5.2e. (Reproduced by courtesy of Philips Components Ltd.)
  b Common emitter equivalent circuit.
  c Common emitter audio amplifier, I_b = base bias or standing current; I_c = collector standing current; i_c = useful signal current in load.
  d Common base RF amplifier.
  e Common collector high-input-impedance audio amplifier.

there is voltage gain also. Figure 5.3d shows a common base stage used as an RF amplifier: the common base configuration is very suitable for this purpose because in a transistor such as the venerable 2N918 or its more modern counterparts, designed specially for use up to UHF, the collector emitter capacitance is very low, resulting in little internal feedback and thus a stable amplifier. However, the maximum gain available from a common base stage is less than for a common emitter stage (stability considerations apart), as the current gain of the device is slightly less than unity. Figure 5.3e shows a common collector stage, often known as an emitter follower. Here, the voltage gain is nearly unity, but there is power gain, as the output impedance of the stage is much lower than its input impedance. It can thus drive a low load impedance without heavily loading the source.

In the early 1960s, the first practical junction field effect transistors made their appearance, though they had been described theoretically as early as 1952. Figure 5.5a shows the symbols for the device while Figure 5.5b and c show the construction and operation of the first type introduced, the depletion mode junction FET or JFET. In this device, in contrast to the bipolar transistor, conduction is by means of majority carriers which flow through the channel between the source (analogous to an emitter) and the drain (analogous to a collector). The gate is a region of silicon of opposite polarity to the source-cum-substrate-cum-drain. When the gate is at the same potential as the source and drain, its depletion region is shallow and current carriers (electrons in the case of the N channel FET shown in Figure 5.5c) can flow between the source and the drain. The FET is thus a unipolar device; minority carriers play no part in its operation. As the gate is made progressively more negative, the depletion region extends across the channel depleting it of carriers, and eventually pinching off the channel entirely when V_{gs} reaches $-V_p$, the pinch-off voltage. Thus for zero or small voltages of either polarity between source and drain, the device can be used as a passive voltage controlled resistor. The JFET is however more normally employed in the active mode as an amplifier (Figure 5.5d) with a positive supply rail (for an N channel FET), much like an NPN transistor stage. Note that even with zero gate/source reverse bias, as the drain becomes more and more positive, the gate becomes negative relative to it, so that the

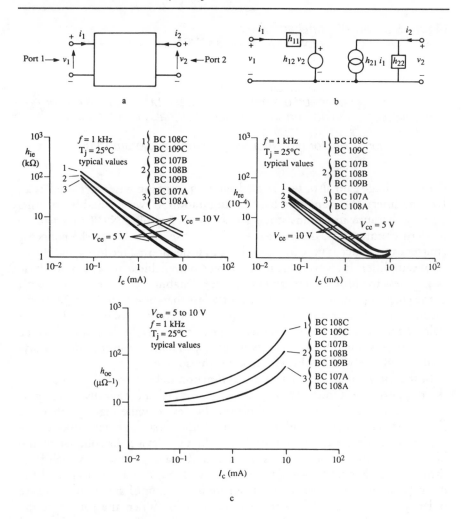

Figure 5.4   h-parameters.

a Generalized two-port black box. v and i are small-signal alternating quantities. At both ports, the current is shown as in phase with the voltage (at least at low frequencies), i.e both ports are considered as resistances (impedances).

b Transistor model using hybrid parameters.

c h-parameters of a typical small-signal transistor family (see also Figure 5.3a). (Reproduced by courtesy of Philips Components Ltd.)

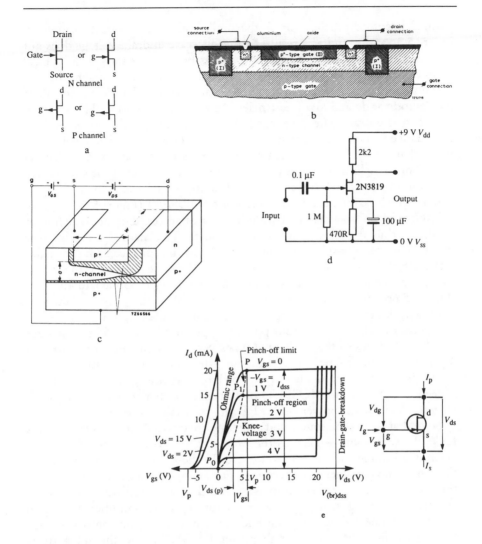

Figure 5.5   *Depletion mode junction field effect transistors.*
   a Symbols.
   b Structure of an N channel JFET.
   c Sectional view of an N channel JFET. The P⁺ upper and lower gate regions should be imagined to be connected in front of the plane of the paper, so that the N channel is surrounded by an annular gate region. The cross-hatched area indicates the pinch-off region.
   d JFET audio-frequency amplifier.
   e Characteristics of N channel JFET; pinch-off voltage $V_p = -6$ V.
(Parts b, c and e reproduced by courtesy of Philips Components Ltd.)

channel becomes pinched off at the drain end. This is clearly shown in Figure 5.5c and e, and as a result, further increase in drain voltage does not increase the drain current appreciably. So as Figure 5.5e shows, the typical drain characteristic is pentode-like. Provided that the gate is reverse biased, as it normally will be, it draws no current, making the FET a close cousin of the pentode at dc and low frequencies. At RF it behaves more like a triode, owing to the drain gate capacitance C_{gd}, analogous to the collector base capacitance of a bipolar transistor. The positive excursions of gate voltage of an N channel FET (or the negative excursions in the case of a P channel device) must be limited to less than 0.5 V to avoid turn-on of the gate/source junction, otherwise the benefit of a high input impedance is lost.

In the metal oxide field effect transistor or MOSFET (Figure 5.6a) the gate is insulated from the channel by a thin layer of silicon dioxide, which is an insulator: thus the gate circuit never conducts. The channel is a thin layer formed between the substrate and the oxide. In the enhancement (normally off) MOSFET, a channel of semiconductor of the same polarity as the source and drain is induced in the substrate by the voltage applied to the gate (Figure 5.6b). In the depletion (normally on) MOSFET, a gate voltage is effectively built in by ions trapped in the gate oxide (Figure 5.6c). Figure 5.6a shows symbols for the four possible types and Figure 5.6d summarizes the characteristics of the N channel types. Since it is much easier to arrange for positive ions to be trapped in the gate oxide than negative ions or electrons, P channel depletion MOSFETs are not generally available. Indeed, for JFETs and MOSFETs of all types, N channel far outnumber P channel devices. RF power MOSFETs are invariably N type.

Note that whilst the source and substrate are internally connected in most MOSFETs, in some — such as the Motorola 2N351 — the substrate connection is brought out on a separate lead. In these cases it is possible to use the substrate as another input terminal. For example, in a frequency changer, the signal could be applied to the gate and the local oscillator (LO) to the substrate, resulting in reduced LO radiation; in an IF amplifier, the signal could be connected to the gate and the automatic gain control voltage (AGC) to the substrate. In high power RF MOSFETs, the substrate is always internally connected to the source.

In the N channel dual-gate MOSFET (Figure 5.7) there is a second gate between gate 1 and the drain. Gate 2 is typically operated at +4 V with respect to the source and serves the same purpose as the screen grid in a tetrode or pentode. It results in a reverse transfer- or feedback-capacitance C_{rss} between drain and gate 1 of only about 0.01 pF, against 1 pF or thereabouts for small-signal JFETs, single-gate MOSFETs and bipolar transistors designed for RF applications. As Figure 5.7c shows, the dual-gate MOSFET is equivalent to a two-transistor amplifier stage consisting of a common source FET driving a common gate FET. It is thus an example of an amplifier known as the cascode stage, which is described in more detail in Chapter 6.

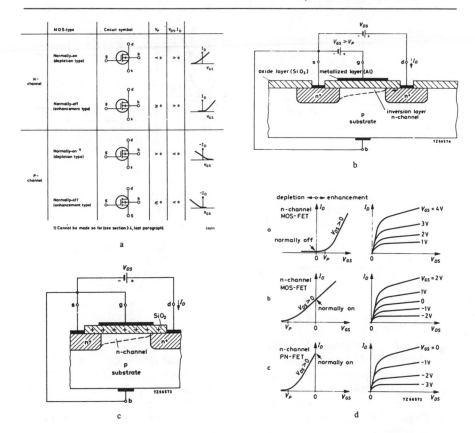

Figure 5.6 *Metal-oxide semiconductor field effect transistors.*
a MOSFET types. Substrate terminal b (bulk) is generally connected to the source, often internally.
b Cross-section through an N channel enhancement (normally off) MOSFET.
c Cross-section through an N channel depletion (normally on) MOS-FET.
d Examples of FET characteristics: (i) normally off (enhancement); (ii) normally on (depletion and enhancement); (iii) pure depletion (JFETs only).
(Reproduced by courtesy of Philips Components Ltd.)

Linearity is an important consideration in amplifiers and other devices for RF applications. This is because a lack of linearity (distortion) can result, in a receiver, in the degradation of a wanted small signal in the presence of large unwanted ones and, in the case of a transmitter, in the unintentional transmission of energy at frequencies other than the authorized transmit frequency, interfering with other users. In an ideal amplifier, the waveform

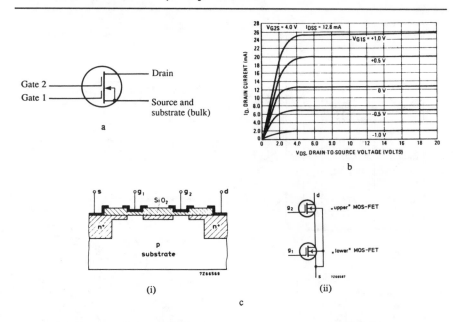

Figure 5.7   *Dual-gate MOSFETs.*
   a Dual-gate N channel MOSFET symbol. Gate protection diodes, not shown, are fabricated on the chip in many device types. These limit the gate/source voltage excursion in either polarity, to protect the thin gate oxide layer from excessive voltages, e.g. static charges.
   b Drain characteristics (3N203/MPF203). (Reproduced by courtesy of Motorola Inc.)
   c Construction and discrete equivalent of a dual-gate N channel MOSFET. (Reproduced by courtesy of Philips Components Ltd.)

of the output is identical to that of the input — only larger. Thus the transfer characteristic of the stage is perfectly linear. There are two main ways in which the characteristic may depart from the ideal. Firstly, the gain may differ on positive- and negative-going half-cycles of the input; Figure 5.8a(i) to (iii) shows how this results in a spurious component in the output at twice the input frequency. This is called second order distortion, since there is an output component proportional to the square of the input voltage. The other common form of distortion is called third order distortion, producing a spurious component in the output at three times the frequency of the input signal. This is illustrated in Figure 5.8b and c, showing what happens when compression of the signal occurs at both positive and negative peaks, due to a cubic or S-shaped component in the transfer characteristic. The top waveform in Figure 5.8c is the amount by which the output falls short of what it would have been had the transfer characteristic been linear. This

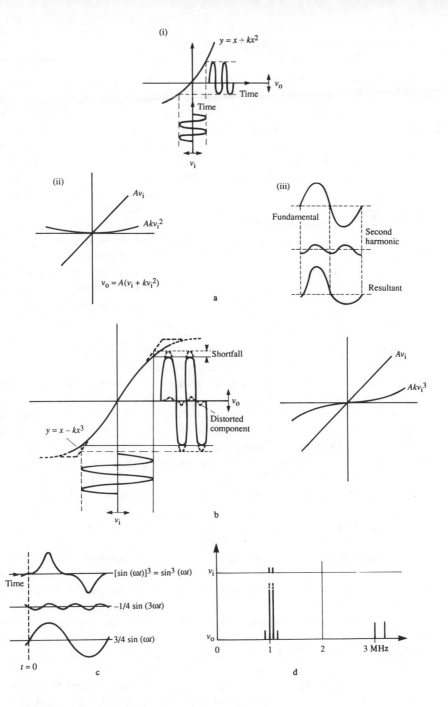

Figure 5.8 *Even-order and odd-order distortion.*
a Second order distortion, typical of a single-ended class A amplifier.
b Third order distortion, typical of a push–pull amplifier.
c Third order distortion analysed.
d Third order intermodulation distortion with two tones of equal amplitude.

shortfall consists of two components, one at ωt representing gain compression, and one at the third harmonic $3\omega t$.

When two signals are present simultaneously, as will commonly happen in the front end of a radio receiver, second-order distortion will also result in products at frequencies equal to the sum and difference of the two input signals. One of these spurious products may fall on top of a small wanted signal, preventing its reception entirely. With third-order distortion, signals at f_1 and f_2 will result in spurious products at $2f_1-f_2$ and $2f_2-f_1$, again possibly jamming a small wanted signal. This is illustrated in Figure 5.8d. Third-order distortion is particularly undesirable, since the spurious products fall close to f_1 and f_2. If f_1, f_2 and the wanted signal are all close together, it will be impossible to provide sufficient selectivity to reduce the amplitude of f_1 and/or f_2 to a level where their third-order intermodulation products are negligible. High linearity is a desirable feature of an active device such as an amplifier, but careful circuit and equipment design is needed if the linearity is to be realized in practice. At the circuit level, linearity is improved by accepting a modest stage gain and possibly including an additional stage, rather than seeking to obtain the maximum possible gain from every stage. Careful attention to layout and screening to avoid feedback (resulting in near instability) is also essential. However carefully designed, there must come a point as the input signal level is increased, where an amplifier overloads. Figure 5.9a shows the input–output relation for an amplifier with a gain of G dB. At low levels, the output rises decibel for decibel with the input, but for very large inputs the amplifier is driven into limiting and reaches its 'saturated output power'. In saturation, there will be a substantial level of harmonic power in the output of the amplifier in addition to the wanted fundamental output, at least in the case of an amplifier stage which does not incorporate a tuned tank circuit. The level at which the fundamental output is 1 dB less than it would be in the absence of limiting is called the compression point.

Figure 5.9b shows that when two fundamentals are applied to an amplifier simultaneously, for low input levels the second-order and third-order intermodulation products are way below the wanted output. Nevertheless, theoretically for every decibel by which the input signals rise, the second-order intermodulation products rise by 2 dB and the third-order products by 3 dB. Empirically, this rule of thumb is found to hold for well-behaved circuits, up to about 10 dB below the compression point. If the results are plotted as in Figure 5.9b and extrapolated, eventually the level of the intermodulation products will notionally intersect the level of the fundamental inputs. The corresponding second- and third-order input intercept points II are shown on the x axis and the output intercept points OI on the y axis. A cheap way for the sharp manufacturer to make his amplifier sound good is to talk a lot about the input intercept points and then just barely mention in passing that the figures he quotes are for the output intercept points.

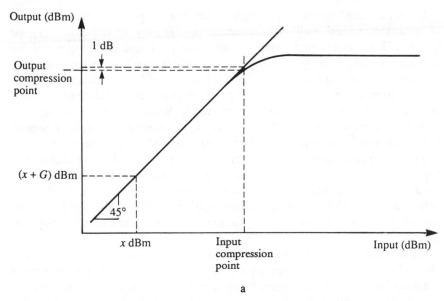

a

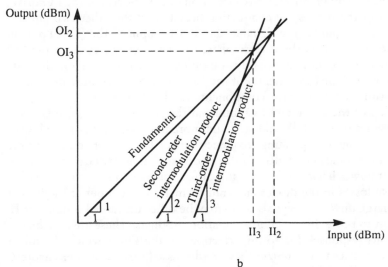

b

Figure 5.9 *Compression and intermodulation.*
a Compression point of an amplifier, mixer or other device with gain G
dB (single tone input).
*b Second- and third-order input and output intercept points (II and IO);
see text (two inputs of equal amplitude).*

Mixers are used to translate a signal from one frequency f_a to another, f_b, by means of a local oscillator frequency f_{LO}. f_b may be either $f_a + f_{LO}$ or $f_a - f_{LO}$. Both active and passive mixers are used and both types will be considered here. A mixer is subject to stringent, not to say contradictory constraints. It is required to exhibit a strong second order characteristic to signals applied to the signal and LO ports, to produce the required sum and difference frequencies, but to be exceedingly linear to two or more large unwanted signals applied to the signal port, in order not to produce second order and more importantly third order intermodulation products. It is also convenient if the mixer is balanced, that is to say that the LO input does not appear at the output port, or alternatively that the signal input does not so appear. A professional communications receiver will usually use a double-balanced mixer (DBM), i.e. one where neither the signal nor the LO input appear at the output, whilst the LO does not appear at the RF input port either.

Figure 5.10a shows on the left the circuit diagram of a typical passive DBM (also known as a ring mixer since all four diodes are connected sequentially anode to cathode), using a matched quad of schottky diodes. On the right is shown the effective circuit on one half-cycle of the LO drive, when two of the diodes are conducting heavily and the other two cut off. The result is to connect the signal at the R (RF) input to the X (IF) port in one phase, and then in the reverse phase on the next half cycle of the LO waveform. The signal is effectively multiplied by $+1$ and -1 on alternate LO half-cycles. The fundamental of the LO and the signal therefore mix to produce sum and difference components at the X port. In practice, the suppression of the signal and LO inputs at the X port in a passive DBM is limited, typically 40–50 dB midband and more like 15–25 dB at the edges of the device's designed operating frequency range. The conversion loss to the signal input is typically 6.5 dB. Of this, 3 dB is inherently due to the split of the output power between the sum and difference frequencies; the rest is due to resistive losses in the diodes and transformers. If the input at the R port includes large unwanted signals there may be other unwanted outputs at IF in addition to those due to intermodulation products. These are all varieties of 'spurious response' due to imperfections in the DBM which the mixer manufacturer tries to minimize: they are discussed further in later chapters. However, the level of spurious responses exhibited by a mixer in practice depends as much if not more upon the user than upon the manufacturer. The spurious responses are minimized when the mixer is run with interfaces having a very low VSWR at all frequencies, at all of its three ports. The manufacturer's published performance data is measured with test gear having a 50 Ω characteristic impedance, usually with a 10-dB 50-Ω pad at each port for good measure. This is quite unrepresentative of actual conditions of use, but it would be impossible to tabulate the performance at all frequencies for all possible combinations of VSWR at the mixer's three

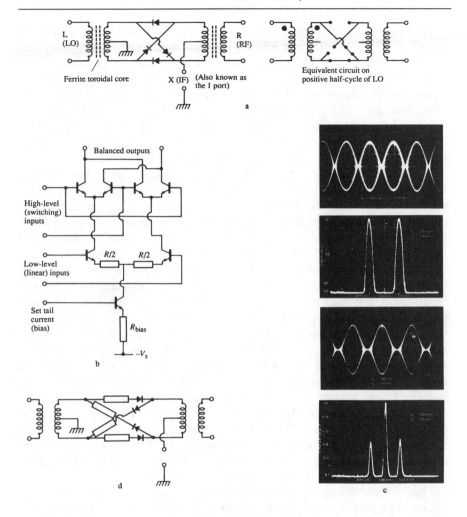

Figure 5.10 *Double-balanced mixers (DBMs).*

a The ring modulator. The frequency range at the R and L ports is limited by the transformers, as also is the upper frequency at the X port. However, the low-frequency response of the X port extends down to 0 Hz (dc).

b Basic seven-transistor tree active double-balanced mixer. Emitter-to-emitter resistance R, in conjunction with the load impedances at the outputs, sets the conversion gain.

c The transistor tree circuit can be used as a demodulator (see text). It can also, as here, be used as a modulator, producing a double-sideband suppressed carrier output if the carrier is nulled, or AM if the null control is offset. The MC1496 includes twin constant current tails for the linear stage, so that the gain setting resistor does not need to be split as in b. (Reproduced by courtesy of Motorola Inc.)

d High dynamic range DBM (see text).

ports. In practice, the mixer's R port is likely to be driven from a low-noise amplifier with a poor output VSWR, or worse still from a band-pass filter, whilst the IF X port is likely to be terminated in a band-pass roofing filter. Pads at the R and X ports are clearly undesirable as they will worsen the receiver's noise figure. A pad at the L port can be useful, albeit at the expense of an increased LO power requirement. A filter connected directly to a mixer port may provide a reasonable match in its pass band, but will reflect energy back into the mixer in its stop band, where its VSWR is very large. Means of avoiding this dilemma are discussed in Chapter 12.

Another well-known scheme, not illustrated here, uses MOSFETs as switches instead of diodes [1]. It is thus, like the Schottky diode ring DBM, a passive mixer, since the MOSFETs are used solely as voltage-controlled switches and not as amplifiers. Reference 2 describes a single balanced active MOSFET mixer providing 16 dB conversion gain and an output third-order intercept point of +45 dBm. Figure 5.10b shows an active DBM of the seven-transistor tree variety; the interconnection arrangement of the four upper transistors is often referred to as a Gilbert cell. The emitter-to-emitter resistance R sets the conversion gain of the stage; the lower its value, the higher the conversion gain but the worse the linearity, i.e. the lower the third-order intercept point. This circuit is available in IC form (see Figure 5.10c) from a number of manufacturers under the type numbers 1496 or 1596, whilst derivatives with a higher dynamic range are also available [3]. Figure 5.10d shows one of the ways the signal handling capability and linearity of the passive DBM can be increased, usually at the expense of a requirement for increased LO drive power. The resistors in series with the diodes swamp and thus stabilize the on resistance of the diodes, whilst the increased forward volt drop increases the reverse bias on the off diodes, minimizing (variations in) their reverse capacitance. High performance DBMs may accept LO drive powers up to +27 dBm.

The term 'active components' for RF must include, in addition to IC mixers, a host of ICs designed to operate as RF or IF amplifier stages, or as complete IF strips, often complete with local oscillator, mixer, and in some cases an RF stage as well. However, the operation of these is so closely bound up with the application circuits, that they are covered in Chapter 6.

References

1. Rafuse, R. P. Symmetric MOSFET mixers of high dynamic range. International Solid State Conference Session XI, University of Pennsylvania, pp. 122–3 (1968)
2. Oxner, E. S. Single balanced active mixer using MOSFETs. In *Power FETs and Their Applications* Prentice Hall, Englewood Cliffs N.J. 07632, p. 292 (1982)
3. Type SL 6440C, GEC Plessey Semiconductors

6

RF small-signal amplifiers

The basic circuit arrangements for a single transistor amplifier stage were described in the last chapter, but there are many practical points of circuit design and these are illustrated below, starting with the common base (common gate) circuit. The low-frequency small-signal input impedance of a grounded base transistor is resistive and equal to $25/I_e$ in ohms, where I_e is in milliamperes. The reciprocal of this gives the mutual conductance g_m, i.e. 40 mA per volt at 1 mA, and pro rata at other collector currents. So, for example, taking a collector current of say 2 mA, gives a grounded base input resistance of 12R5, and this may be taken as a starting point for circuit design even at higher frequencies, in the absence of more specific data. It is too low an impedance to connect directly to an aerial input, so the grounded base amplifier of Figure 6.1a, designed for the VHF FM band, uses a 2:1 turns ratio transformer to match from 12R5 up to 50 Ω. Of course, for more precise circuit design one could measure with a network analyser the actual input impedance of the device at the intended frequency of operation and collector current. However, in the absence of a network analyser, the rough and ready estimate may be used and will result in only a small loss of stage gain compared with a more exact approach. Alternatively, a fairly exact circuit design can be effected using an RF oriented CAD (computer aided design) package, which would probably have a model of the transistor to be used in its component library. The results of the simulation will give a fair idea of the performance to be expected from the hardware as built, provided great care is taken in the practical layout to avoid introducing parasitic capacitive and inductive elements which do not appear in the circuit as modelled.

In grounded base the current gain is less than unity, so the circuit stage gain in Figure 6.1a is explained by the fact that the collector circuit impedance is around 200 Ω (assuming a 50 Ω load at PL_1), at least if the two halves of the output tuned inductor are closely coupled, so that it acts as a 2:1 step-down transformer. Since power equals I^2R and the signal current is (almost) the same in the collector circuit as in the emitter circuit, the power

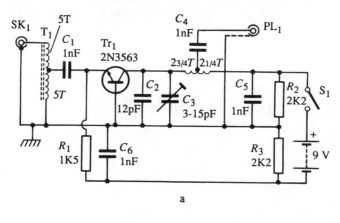

a

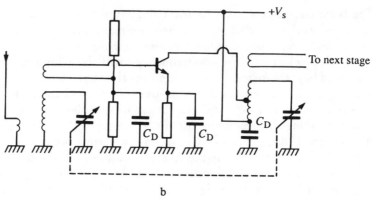

b

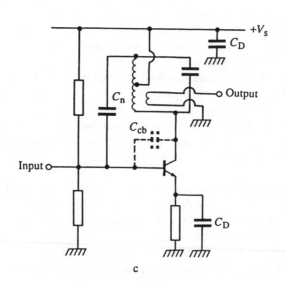

c

Figure 6.1   *RF amplifier stages.*

a Common base RF amplifier with aperiodic (broadband) input and tuned output stages (Reproduced from 'VHF preamplifier for band II', *Ian Hickman, Practical Wireless, June 1982, p. 68, by courtesy of* Practical Wireless.*)*

b Common emitter RF amplifier stage with both input and output circuits tuned. C$_D$ are decoupling capacitors.

c Bridge neutralization. The internal feedback path is not an ideal capacitor C$_{cb}$ as shown, but will have an in-phase component also. If the phase angle of the neutralization via C$_n$ is adjusted, e.g. by means of an appropriate series resistance, the neutralization is more exact – at that particular frequency. The stage is then described as 'unilateralized' at that frequency.

gain is just 200 Ω/12R5 or 16 times, and 10 log(16) equals 12 dB. Of course this approximate approach to circuit design ignores a number of factors; it assumes that the output conductance of the stage is low compared with (1/200) Ω^{-1}, or 0.005 S, the Siemen being the name for the unit of conductance (the output susceptance is absorbed into the tuned circuit). It also ignores the effect of less than unity coupling between the two halves of the output inductor and the effect of internal feedback inside the transistor. This will slightly reduce the stability margin; more so if the stage is used with a 75 Ω source and load, as would in practice be the case. That these factors can indeed by largely ignored in this case, at least to a first approximation, is demonstrated by the measured gain of the circuit which was 11 dB in a 50 Ω system — a very fair agreement with the predicted 12 dB, for a design method involving no more than simple mental arithmetic. A grounded gate FET could alternatively be used in the circuit, and if one were available with a mutual conductance of 20 mA per volt, it would provide a direct match to 50 Ω without needing T$_1$. However, the g$_m$ in a typical small-signal RF FET would be lower than this, so the stage tain would be lower too. If greater selectivity than that provided by the single tuned circuit in Figure 6.1a were required, the transistor's input transformer could be replaced by a tuned circuit with a tap for the antenna input and a coupling coil to the device's emitter.

The common emitter stage potentially provides a greater stage gain than the common base, provided that the gain can be realized, having due regard for stability considerations. Figure 6.1b shows a bipolar common emitter amplifier stage with input and output both tuned. This is an arrangement that might be used for the input stage of an HF communications receiver covering 2–30 MHz; it enables one to provide more selectivity than could be achieved with only one tuned circuit, whilst avoiding some of the complications of coupled tuned circuits. The latter can provide a better band-pass shape — in particular a flatter pass band — but for a communications

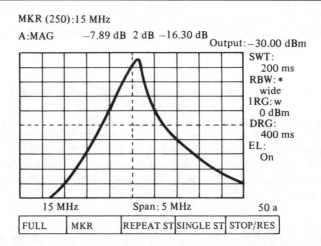

MKR (250):15 MHz
A:MAG   −7.89 dB   2 dB −16.30 dB
Output: −30.00 dBm

SWT:
    200 ms
RBW: *
    wide
1RG: w
    0 dBm
DRG:
    400 ms
EL:
    On

15 MHz   Span: 5 MHz   50 a

| FULL | MKR | REPEAT ST | SINGLE ST | STOP/RES |

Figure 6.2 *Frequency response of an amplifier with unintentional internal feedback. Gain falls faster on the high-frequency side of the peak. Measured using the network analyser illustrated in Figure 14.4.*

receiver covering 2–30 MHz, two single tuned circuits as in Figure 6.1b provide an adequate pass band in any case. With the continued heavy usage of the HF band, RF stages (with the front-end selectivity they can provide) are coming back into favour again. However, an RF amplifier with both input and output circuits tuned needs very careful design to ensure stability, especially when using the common emitter configuration. The potential source of trouble is the collector/base capacitance, which provides a path by which energy from the output tuned circuit can be fed back to the base input circuit. The common emitter stage provides inverting gain, so that the output is effectively 180° out of phase with the input. The current fed back through the collector/base capacitance will of course lead the collector/base voltage by 90°. At a frequency somewhat below resonance (see the Universal Resonance Curve, Appendix 4) the collector voltage will lead the collector current, and the feedback current via the collector/base capacitance will produce a leading voltage across the input tuned circuit. At the frequency where the lead in each tuned circuit equals 45°, there is thus a total of 180° of lead, cancelling out the inherent phase reversal of the stage and leaving us with positive feedback. The higher the stage gain and the higher the Q of the tuned circuits, the more likely the feedback is to cause oscillation, since when the phase shift in each tuned circuit is 45°, its amplitude response is only 3 dB down (see Appendix 4). Even if oscillation does not result, the stage may show a much faster rate of fall of gain to a signal with detuning on the high frequency side than on the lower. This is a sure sign of significant internal feedback (Figure 6.2): with further detuning, the rate of fall of gain approaches 12 dB/octave on both sides of the peak — it only looks faster on the low-frequency side in the figure because the horizontal frequency axis is linear, not logarithmic.

A common technique for increasing the stability margin of an RF amplifier — it could be applied to the circuit of Figure 6.1b — is mismatching. This simply means accepting a stage gain less than the maximum that could be achieved in the absence of feedback. In particular, if the collector (or drain) load is reduced, the stage will have a lower voltage gain. So the voltage available to drive current through the feedback capacitance C_{cb} is reduced pro rata. Likewise, if the source impedance seen by the base (or gate) is reduced, the current fed back will produce less voltage drop across the input circuit. Both measures reduce gain and increase stability: the gain sacrificed by mismatching may be recovered by adding another amplifier stage. This may be a cheaper solution than obtaining the required gain from fewer stages by adding circuit complexity such as 'unilateralization'. This cumbersome term is used to indicate any scheme that will reduce the effective internal feedback in an amplifier stage, i.e. to make the signal flow in the forward direction only. Data sheets for RF devices often quote a figure for the maximum available gain at a given frequency (MAG) and a higher figure for the maximum unilateralized gain (MUG). The traditional term for unilateralization is neutralization, though the latter usually only compensates for the reactive component of the feedback path, whereas the former allows for a resistive component as well. Figure 6.1c shows one popular neutralization scheme, sometimes known as bridge neutralization. The output tuned circuit is centre tapped so that the voltage at the top end of the inductor is equal in amplitude to, and in antiphase to, the collector voltage. The neutralizing capacitor C_n has the same value as the typical value of the transistor's C_{cb}, or it can be a trimmer capacitance set to the same value as the C_{cb} of the individual transistor. The criterion for setting the trimmer is that the response of the stage about the tuned frequency should be symmetrical. This occurs when there is no net feedback, either positive or negative. The series capacitance of C_n and C_{cb} appears across the output tuned circuit and is absorbed into its tuning capacitance, whilst the parallel capacitance of C_n and C_{cb} appears across the input tuned circuit and is absorbed into its tuning capacitance. Neutralization can be very effective for a small-signal amplifier, but is less so for a stage handling large signals. This is because the feedback capacitance C_{cb}, being due to a reverse biased semiconductor junction, varies with the reverse voltage and for large signal swings is thus non-linear.

The common collector circuit (emitter follower) is also useful at RF, mainly as a buffer stage, untuned or at least only tuned at the input. However, be warned that the emitter follower has a reputation for instability unless care is taken in the layout and decoupling of the stage. The internal capacitances C_{ce} and C_{be} can turn the circuit into a Colpitts oscillator (see Chapter 7) in conjunction with any inductance in the base and collector leads. With all three of the basic single transistor stages offering the possibility of instability due to internal feedback, a useful circuit in many applications is the two transistor 'cascode' amplifier stage, which inherently has very little feedback from output to input (Figure 6.3a). The input transistor is used in the grounded emitter configuration, which provides much more current gain than grounded base, whilst also having a higher input impedance. However, there is no significant feedback from the collector circuit to the base tuned circuit, since the collector load of the input transistor consists of the very low emitter input impedance of the second transistor. This is used in the grounded base configuration, which again results in very low feedback from its output to its input. With a suitable transistor type, the cascode can provide well over 20 dB of gain at 100 MHz together with a reverse isolation of 70 dB. This makes it an ideal buffer stage between the VCO of a synthesizer and the variable ratio divider or

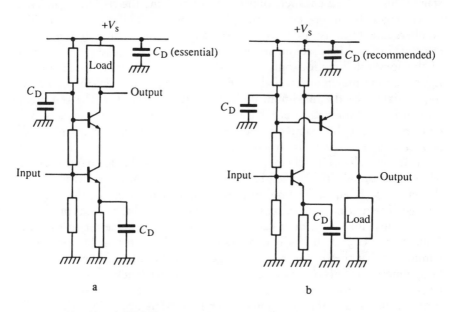

a   b

60, 105 and 200 MHz power gain and noise figure test circuit

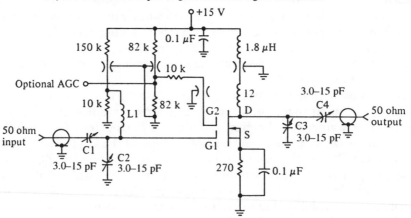

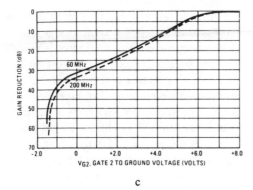

c

Figure 6.3 *Variations on the cascode amplifier.*
a Cascode amplifier.
b Complementary cascode. The load may be a resistor, an RL
combination (peaking circuit), a tuned circuit or a wide band RF trans-
former. C_D are decoupling capacitors.
*c Dual-gate MOSFET VHF amplifier with AGC, with gain reduction
curve. Maximum gain 27 (20) db at 60 (200) MHz with no gain reduction
(V_{g2} at +7.5 V). The Motorola MPF131 provides an AGC range featuring
up to 60 dB of gain reduction. (Reproduced by courtesy of Motorola Inc.)*

two-modulus prescaler, removing the possibility of comparison frequency
sidebands in the synthesizer's output caused by dynamic variations of the
divider's input capacitance. Figure 6.3b shows an interesting variation on the
theme. Here, the grounded base stage uses a PNP transistor. The result is
that the output is ground-referenced, with no RF current drawn from the
positive rail, easing decoupling requirements. Figure 6.3c shows a cascode
stage in a single device, using a semiconductor tetrode or dual-gate
MOSFET. In addition to a 2.5 dB noise figure and a stable forward gain of
27 (20) dB at 60 (200) MHz, it provides an AGC capability with up to 60 dB
of gain reduction.

Reverse isolation is an important parameter of any RF amplifier and is
simply determined by measuring the 'gain' of the circuit when connected
back to front, i.e. with the input applied to its output port and the output
taken from its input. This is easily done in the case of a stand-alone amplifier
module, but not so easy when the amplifier is embedded in a string of
circuitry in an equipment. In the days of valves one could easily derive a
stage's reverse isolation (knowing its forward gain beforehand) simply by
disconnecting one of the heater leads and seeing how much the gain fell.
When a valve is cold it provides no amplification, so signals can pass only via
the inter-electrode capacitances, and these are virtually the same whether the

valve is hot or cold. With no gain provided by the valve, the forward and reverse isolation are the same. Much the same dodge could be used with transistors, by open circuiting the emitter to dc but leaving it connected as before at ac. However, the results are not nearly so reliable as in the valve case, as many of the transistor's parasitic reactances will change substantially when the emitter current is reduced to zero. For an RF amplifier to be stable, clearly its reverse isolation should exceed its forward gain by a reasonable margin, which need not be anything like the 40–80 dB obtainable with the cascode mentioned above. A difference of 20 dB is fine and 10 dB adequate, whilst some commercially-available broadband RF amplifier modules quote a reverse isolation which falls to as little as 3 dB in excess of the forward gain at the top of their frequency range.

In the early stages of a radio receiver, an amplifier may be subjected at its input to large unwanted signals in addition to the wanted signal. To prevent any resultant degradation of the wanted signal, the amplifier must possess high linearity; this topic is covered in Chapter 5. However, linearity is only one of several very important qualities of an input amplifier stage. It must also exhibit a low noise figure and a high dynamic range. The silicon atoms of the atomic lattice which constitutes the transistor are in a state of 'thermal agitation' which is proportional to the absolute temperature. Consequently the flow of carriers through the transistor is not smooth and orderly but noisy, like the rushing of a mountain stream. Like the noise of the stream, no one frequency predominates. Electrical noise of this sort is called thermal agitation noise, or just thermal noise, and its intensity is independent of frequency (or 'white') for most practical purposes. The available noise power associated with a resistor is independent of its resistance and is equal to -174 dBm/Hz, e.g. in a 3 kHz communications bandwidth, to -139 dB relative to a level of 1 mW. This means that the wider the bandwidth we consider, the higher the noise power it contains. It seems that if we consider an infinite bandwidth, there would be an infinite amount of power available from a resistor, but in fact, the noise bandwidth is inherently limited; at room temperature thermal noise starts to tail off beyond 1000 GHz (10% down), the noise density falling to 50% at 7500 GHz (Figure 6.4b). At very low temperatures such as are used with maser amplifiers, e.g. 1 K ($-272°$C), the noise density is already 10% down at 5 GHz.

Returning to our RF amplifier then, if it is driven from a 50R source there will be noise power fed into its input therefrom (Figure 6.4a). If the amplifier is matched to the source, i.e. its input impedance is 50 Ω resistive, the rms noise voltage at the amplifier's imput v_n is equal to half the source resistor's open-circuit noise voltage, i.e. to $\sqrt{(kTRB)}$, where R is 50 Ω, k is Boltzmann's constant $= 1.3803 \times 10^{-23}$ J/K and B is the bandwidth of interest. At a temperature of 290K (17°C or roughly room temperature) this works out at 24.6 nV in 50 Ω in a 3 kHz bandwidth. If the amplifier were perfectly noise-free and had a gain of 20 dB (i.e. a voltage gain of $\times$ 10,

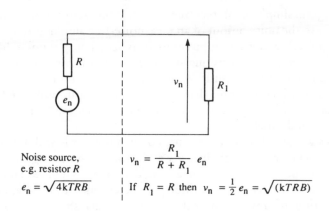

$$v_n = \frac{R_1}{R + R_1} e_n$$

Noise source, e.g. resistor R

$$e_n = \sqrt{4kTRB}$$

If $R_1 = R$ then $v_n = \frac{1}{2} e_n = \sqrt{(kTRB)}$

a

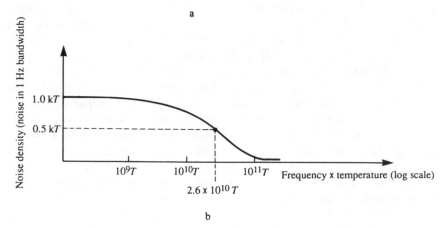

b

Figure 6.4 *Thermal noise.*

a A noisy source such as a resistor can be represented by a noise-free resistor R of the same resistance, in series with a noise voltage generator of EMF e_n = √(4kTRB) volts. Available noise power = v_n^2/R = $(e_n/2)^2/R$ = P_n say. At room temperature (290 K) p_n = –204 dBW in a 1 Hz bandwidth = –174 dBm in a 1 Hz bandwidth. If B = 3000 Hz then P_n = –139 dBm, and if R = R_1 = 50 Ω then v_n = 0.0246 μV in 3 kHz bandwidth.

b Thermal noise is 'white' for all practical purposes. The available noise power density falls to 50% at a frequency of 2.6 × 10^{10} T, i.e. at about 8000 GHz at room temperature, or 26 GHz at T = 1 K.

assuming its output impedance is also 50 Ω), we would expect 0.246 μV rms noise at its output: if the output noise voltage were twice this, 0.492 μV rms, we would describe the amplifier as having a noise figure of 6 dB. Thus the noise figure simply expresses the ratio of the actual noise output of an amplifier to the noise output of an ideal noise-free amplifier of the same gain. The amplifier's equivalent input noise is its actual output noise divided by its gain. Chapter 5 also introduced the concept of compression level. The dynamic range of an amplifier simply means the ratio between the smallest input signal which is larger than the equivalent input noise, and the largest input signal which produces an output below the compression level, expressed in decibels.

The catalogue of desirable features of an amplifier is still not complete; in addition to low noise, high linearity and wide dynamic range, the gain, input impedance and output impedance should all be well defined and repeatable. Further, steps to define these three parameters should, ideally, not result in deterioration of any of the others. Figure 6.5a shows a broadband RF amplifier with its gain, input and output impedance determined by negative feedback [1]. The resistors used in the base and emitter feedback circuits necessarily contribute some additional noise. This can be avoided by the scheme known as lossless feedback [2] shown in Figure 6.5b. Here the gain, input and output impedances are all determined by the ampere-turn ratios of the windings of the transformer. This arrangement results in a very low noise figure, but the reverse isolation of the stage is unfortunately low.

In the later stages of a receiver, the requirement for a very low noise figure may be somewhat relaxed, whilst band-pass filtering preceding the IF stages prevents large unwanted signals reaching them, relaxing linearity and dynamic range requirements (as is covered more fully in Chapter 10). This easing of the requirements has led to discrete transistor IF stages giving way to integrated circuits purpose-designed to provide stable gain and a wide range AGC capability. IC RF amplifiers are also used in the less demanding RF amplifier applications, for instance in a transmitter exciter, where the signal to be transmitted is the only signal. A typical range of such ICs is the GEC Plessey Semiconductors SL600/6000 series of devices, the SL610C and SL611C being RF amplifiers and the SL612C an IF amplifier. These devices provide 20–34 dB gain according to type, and a 50 dB AGC range. In FM receivers, the amplitude of the received signal conveys no information, so a limiting IF strip can be used. This typically has a number of amplifier stages in cascade. Here, with a minimum level input signal there is just enough gain to drive the last stage into saturation or 'limiting', whilst as the signal level increases, more and more stages operate in limiting, each being designed to overload cleanly and to accept an input as large as its saturated output. A popular example is the CA3189 available from a number of manufacturers, it is an improved performance replacement for the earlier CA3089. With three limiting stages it provides a typical 10.7 MHz sensitivity of 10 µV for limiting, and includes a double balanced quadrature detector (for use with external quadrature coil), audio amplifier with muting circuit, and provides AFC and delayed AGC outputs for the tuner.

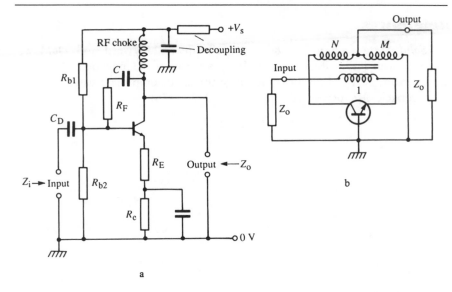

Figure 6.5 *Input and output impedance determining arrangements.*

a Gain, input and output impedances determined by resistive feedback. R_{b1}, R_{b2} and R_e determine the stage dc conditions. Assuming the current gain of the transistor is 10 at the required operating frequency, then for input and output impedances in the region of 50 Ω, $R_F = 50^2/R_E$. For example, if $R_E = 10\ \Omega$, $R_F = 250\ \Omega$, then $Z_i \approx 35\ \Omega$, $Z_0 \approx 65\ \Omega$ and stage gain ≈ 10 dB, while if $R_E = 4.7\ \Omega$, $R_F = 470\ \Omega$, then $Z_i \approx 25\ \Omega$, $Z_0 \approx 95\ \Omega$ and gain ≈ 15 dB. C_D are blocking capacitors, e.g. 0.1 μF.

b Gain, input and output impedances determined by lossless (transformer) feedback. The absence of resistive feedback components results in a lower noise figure and higher compression and third order intercept points. Under certain simplifying assumptions, a two-way match to Z_0 results if $N = M^2 - M - 1$. Then power gain $= M^2$, impedance seen by emitter $= 2Z_0$ and by the collector $= (N + M)Z_0$. This circuit arrangement is used in various broadband RF amplifier modules produced by Anzac Electronics Division of Adams Russel and is protected by US Patent 3 891 934: 1975 (dc biasing arrangements not shown). (Reprinted by permission of Microwave Journal.)

Numerous special purpose IC amplifiers for RF and/or IF applications are available from a number of specialist manufacturers, e.g. Avantec, GEC Plessey Semiconductors, Mini Circuits Laboratories, Motorola and others. The products offered include low phase shift limiters for phase recovery strips in radar and ECM systems, multistage log/limiting amplifiers with IF and video outputs for radar receivers, low power IF strips with PLL detector and squelch outputs for narrow-band FM communications, etc.

References

1. *Solid State Design for the Radio Amateur.* Hayward and DeMaw, American Radio Relay League Inc., Newington, Connecticut, USA
2. Norton, D. E. High dynamic range transistor amplifiers using lossless feedback. *Microwave Journal*, May, 53–7 (1976)

7
Modulation and demodulation

Modulation is the process of impressing information to be transmitted onto an RF 'carrier' wave, in such a way that it can be retrieved again in more or less undistorted form at the receiver. Figure 7.1a shows how information is transmitted by CW (continuous wave) using the Morse code, once widely used on the HF band (1.6–30 MHz) for commercial marine traffic and still used by amateurs for world-wide DX-ing on a few watts. Broadcasting on the long, medium and short wavebands uses AM (amplitude modulation) (Figure 7.1b). The amplitude of the RF carrier wave changes to reflect the instantaneous value of the modulating baseband waveform, e.g. speech or music. The baseband signal is limited to 4.5 kHz bandwidth, restricting the bandwidth occupied by the transmitted signal to 9 kHz, centred on the carrier frequency. With maximum modulation by a single sinusoidal tone, the transmitted power is 50% greater than with no modulation; this is the 100% modulation case. Note that the power of the carrier is unchanged, so that at best only one-third of the transmitted power is used to convey the baseband information — even less during average programme material. For this reason, single sideband (SSB) modulation has become very popular with military, commercial and amateur users for voice communication at HF. In SSB (Figure 7.1c), only one of the two sidebands is transmitted, the other and the carrier being suppressed. Spectrum occupancy is halved and all transmitted power is useful information. At the receiver, the missing carrier must be supplied by a carrier re-insertion oscillator at exactly the appropriate frequency; an error of up to 10 Hz or so is acceptable on speech, less than 1 Hz on music. In the early days of SSB this was difficult and a very fine tuning control called a clarifier was provided, but with synthesized transmitters and receivers this is no longer a problem. In commercial and military SSB applications USB (upper sideband) operation is the norm, in amateur practice USB is used above 10 MHz and LSB below. ISB (independent sideband) operation is occasionally used commercially. Here, one communication channel is carried on the lower sideband and an entirely

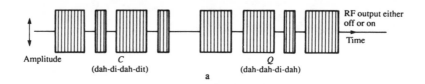

Amplitude

C
(dah-di-dah-dit)

Q
(dah-dah-di-dah)

RF output either
off or on

Time

a

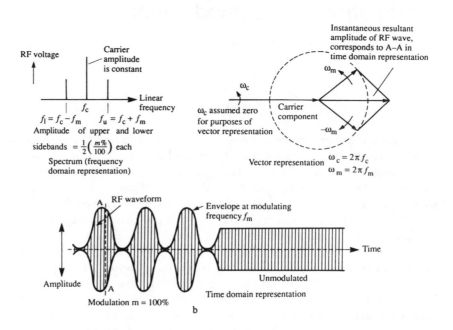

RF voltage

Carrier
amplitude
is constant

f_c

$f_l = f_c - f_m$ $f_u = f_c + f_m$

Amplitude   of upper   and lower

sidebands $= \frac{1}{2}\left(\frac{m\%}{100}\right)$ each

Spectrum (frequency
domain representation)

Linear
frequency

ω_c assumed zero
for purposes of
vector representation

Carrier
component

Instantaneous resultant
amplitude of RF wave,
corresponds to A–A in
time domain representation

ω_m

$-\omega_m$

Vector representation $\omega_c = 2\pi f_c$
$\omega_m = 2\pi f_m$

A   RF waveform

Envelope at modulating
frequency f_m

Time

Amplitude

A

Modulation m = 100%

Unmodulated

Time domain representation

b

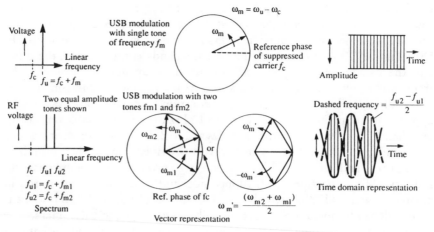

Voltage

Linear
frequency

f_c
$f_u = f_c + f_m$

USB modulation
with single tone
of frequency f_m

$\omega_m = \omega_u - \omega_c$

ω_m

Reference phase
of suppressed
carrier f_c

Amplitude

Time

RF
voltage

Two equal amplitude
tones shown

USB modulation with two
tones fm1 and fm2

ω_{m2}

ω_m'

ω_m'

Dashed frequency $= \dfrac{f_{u2} - f_{u1}}{2}$

Linear frequency

f_c f_{u1} f_{u2}

$f_{u1} = f_c + f_{m1}$
$f_{u2} = f_c + f_{m2}$

Spectrum

or

ω_{m1}

$-\omega_m'$

Ref. phase of fc

$\omega_m' = \dfrac{(\omega_{m2} + \omega_{m1})}{2}$

Vector representation

Time

Time domain representation

c

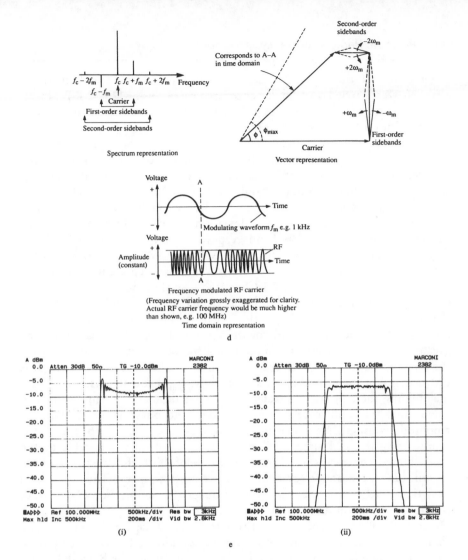

Spectrum representation

Vector representation

Time domain representation

d

(i)   (ii)

e

Figure 7.1 *Types of modulation of radio waves.*

a CW (ICW) modulation. The letters CQ in Morse (seek you?) are used by amateurs to invite a response from any other amateur on the band, to set up a QSO (Morse conversation).

b AM: 100% modulation by a single sinusoidal tone shown.

c SSB (USB) modulation. Note that with two-tone modulation, the signal is indistinguishable from a double-sideband suppressed carrier signal with a suppressed carrier frequency of $(f_{u1} + f_{u2})/2$. This can be seen by subtracting the carrier component from the 100% AM signal in b. The upper and lower halves of the envelope will then overlap as in c, with the RF phase alternating between 0° and 180° in successive lobes.

d FM. For maximum resultant phase deviation ϕ up to about 60° as shown, third- and higher-order sidebands are insignificant.

e Power spectral density (PSD), very wide band FM with (i) sinewave and (ii) triangular modulation. Note: envelope of PSD is shown. The areas are filled with discrete lines spaced at the frequency of the modulating waveform, f_m. Fall-off beyond $\pm f_{dmax}$ is rapid.

different one on the upper. At one time, four international telephone trunk channels were carried on a single suppressed carrier using '2 + 2 ISB'. Here, each sideband carried two telephone channels, one at baseband and one translated up to the band 4–8 kHz.

Figure 7.1d illustrates frequency modulation. FM was proposed as a modulation method even before the establishment of an AM broadcasting service, but it was not pursued as the analysis showed that it produced sidebands exceeding greatly the bandwidth of the baseband signal [1]. FM is used for high fidelity broadcasting in the internationally allocated VHF FM band 88–108 MHz, using a peak deviation of ±75 kHz around the RF carrier frequency and a baseband response covering 50 Hz to 15 kHz. Figure 7.1 shows the characteristics of AM and FM in three ways: in the frequency domain, in the time domain and as represented in vector diagrams. Note that in Figure 7.1d a very low level of modulation is shown, corresponding to a low amplitude of the baseband modulating sinewave (frequency f_m). Even so, it is clear that if only the sidebands at the modulating frequency existed, the amplitude of the RF signal would be greatest twice per cycle of the modulating frequency, at the instants when the phase deviation of the RF from the unmodulated state was greatest. It is the presence of the second order sidebands at $2f_m$ that compensates for this, maintaining the amplitude constant. At wider deviations, many more FM sidebands appear, all so related in amplitude and phase as to maintain the amplitude constant. Note that the maximum phase deviation of the vector representing the FM signal will occur at the end of a half-cycle of the modulating frequency, since during the whole of this half-cycle the frequency will have been above (or below) the centre frequency. Thus the phase deviation is 90° out of phase with the frequency deviation. For a given peak frequency deviation, the peak phase deviation is inversely proportional to the modulating frequency, as is readily shown. Imagine the modulating signal is a 100 Hz squarewave and the peak deviation is 1 kHz. Then during the 10 ms occupied by a single cycle of the modulation, the RF will be first 1000 Hz higher in frequency than the nominal carrier frequency and then, during the second 5 ms, 1000 Hz lower. So the phase of the RF will first advance steadily by five complete cycles (or 10π rad) and then crank back again by the same amount; i.e. the peak phase deviation is $\pm5\pi$ rad relative to the phase of the unmodulated carrier. Now the average value of a half-cycle of a sinewave is $2/\pi$ times that of a half-cycle of a squarewave of the same peak amplitude; so if the modulating signal had been a sinewave, the peak phase deviation would have been just ±10 rad. Note that the peak phase deviation in radians (for sinewave modulation) is just f_d/f_m, the peak frequency deviation divided by the modulating frequency: this is known as the modulation index of an FM signal. If the modulating frequency had been 200 Hz (and the peak deviation 1 kHz as before), the shorter period of the modulating frequency would result in the peak-to-peak phase change being

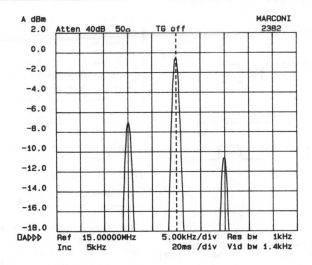

Figure 7.2 *15 MHz carrier with both FM and AM sidebands.*

halved to ±5 rad; so for a given peak frequency deviation, the peak phase deviation is inversely proportional to the modulating frequency.

For monophonic FM broadcasting the peak frequency deviation is ±75 kHz, so the peak phase deviation corresponding to 100% sinewave modulation would be ±5 rad at 15 kHz and ±1500 rad at 50 Hz modulating frequency. Thus on reception, 1 rad of spurious deviation at 50 Hz due to noise will have much less effect than 1 rad of deviation at 15 kHz, giving rise to the well-known triangular noise susceptibility of FM. It also explains the greater signal to noise ratio required for stereo reception, since the left minus right difference signal is a 15 kHz double sideband signal occupying the spectrum 23–53 kHz, modulated on a suppressed 38 kHz sub-carrier. Quite apart from the slightly wider IF bandwidth compared with mono needed to receive stereo FM transmissions, the difference signal is inherently more susceptible to noise degradation as indicated by the triangular noise susceptibility characteristic of FM reception. The noise susceptibility in the upper part of the baseband mono compatible sum signal is reduced by applying a 6 dB per octave pre-emphasis above 3.2 kHz, which effectively produces PM (phase modulation) at the higher audio frequencies. A corresponding de-emphasis is applied in the receiver. The pre-emphasis breakpoint corresponds to a time constant of 50 μs (2.1 kHz and 75 μs are the values used in the USA).

If the modulation index is small compared with unity, the second and higher order sidebands are negligible, but if it is very much larger than unity there are a large number of significant sidebands and these occupy a

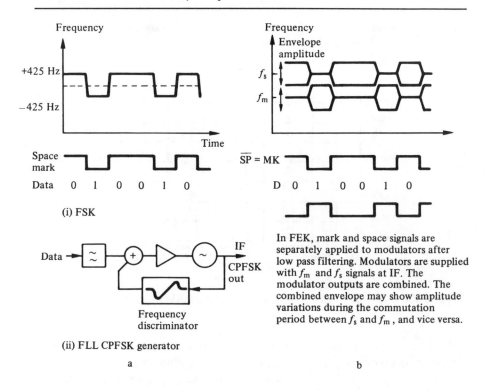

(i) FSK

(ii) FLL CPFSK generator

In FEK, mark and space signals are separately applied to modulators after low pass filtering. Modulators are supplied with f_m and f_s signals at IF. The modulator outputs are combined. The combined envelope may show amplitude variations during the commutation period between f_s and f_m, and vice versa.

a   b

Figure 7.3   *Two methods of modulating a carrier with digital data.*
   a FSK.
   b FEK.

bandwidth virtually equal to $2f_d$, i.e. the bandwidth over which the signal sweeps. The usual approximation for the bandwidth of an FM signal is BW = $2(f_d + f_m)$. Note that if one of the first-order FM sidebands in Figure 7.1d were reversed, they would look exactly like a pair of AM sidebands; this is why one of the first-order FM sidebands in the frequency domain representation has been shown inverted. A spectrum analyser is not sensitive to the relative phases of the signals it encounters during its sweep, so it will show the carrier and sidebands of an AM or of a low-deviation FM signal as identical. However, if the first-order sidebands displayed are unequal in amplitude, this indicates that there is both amplitude and frequency modulation present on the carrier; this is illustrated in Figure 7.2. Figure 7.1e shows the spectra of high modulation index FM for both sinewave and triangular wave modulation with a frequency f_m. In both cases, the overall shape of the power distribution versus frequency is shown. It consists of discrete spectral lines spaced at intervals f_m, with an overall envelope the same shape as the power density plot of the modulating waveform. The flat power density plot with triangular modulation is useful in a jammer application and a very high modulation index ensures a rapid fall away in power outside the intentionally jammed band, avoiding interference with own communications. However, to jam a bandwidth of many megahertz with lines close enough to ensure jamming even a narrow band target, will require a low modulating frequency. This means that the 'revisit time' for a channel, especially one near the edge of the jammed bandwidth, may become overlong. A narrow band of noise may therefore be added to a rather higher frequency triangular wave modulating signal, to spread out the modulation, filling in the gaps between spectral lines.

Many modulation methods have been employed for the transmission of digital data, or of information in digital form such as teleprinter traffic. They are all variations of AM, FM or PM, or of a combination of these. One of the earliest is FSK (frequency shift keying) which is widely used for the transmission of text in ITA2 (international teleprinter alphabet No. 2) by national news agencies (see Figure 7.3a). A commonly used standard on HF is 850 Hz shift (±425 Hz on the suppressed carrier frequency). If the change from one frequency, representing a zero, to the other, representing a one, is abrupt, then the signal will occupy a greater bandwidth than is necessary for its successful reception: the excessive OBW (occupied bandwidth) may interfere with other stations. Several means are used to avoid this, such as band-pass filtering the FSK signal in the exciter before passing it to the PA (power amplifier), shaping or low-pass filtering the data stream and its inverse before applying to two amplitude modulators (this method is known as FEK, frequency exchange keying — Figure 7.3b) or generating the FSK signal by feeding the data stream into an FLL (frequency lock loop). In this latter method, there are no phase discontinuities so it is known as CPFSK (continuous phase FSK). Typically, the transition is arranged to occupy about 10% of a bit period and the data rate with 850 Hz shift would usually be 50 baud.

The baud is the unit of signalling rate over the communications link, and the useful bit rate may be lower or higher than this. For example, in ITA2, each character of the message is transmitted as a start bit followed by five data bits followed by one and a half stop bits, giving a bit rate of two-thirds of the baud rate — or rather less in practice. As the code incorporates start and stop bits it operates asynchronously; one character does not need to follow the next immediately, it can dwell on a stop bit until the next character arrives, e.g. from a typist at a keyboard. The five data bits permit 32 different characters to be encoded, so that figure shift and letter shift characters are used to accommodate the alphabet (capitals only), numerals, punctuation and control symbols. ASCII code (American Standard Code for the Interchange of Information, also known as ITA5) uses seven data bits per character giving 128 possibilities and so can support upper and lower case, without needing shift characters. Often an eighth bit is added for parity, a character thus occupying exactly one byte, and many modems accommodate data with one, one and a half or two stop bits — so there may be up to eleven bits to a character.

FSK/FEK may be very simply demodulated using a frequency discriminator and this was originally the usual method, but it is not optimum. A better scheme is to make use of the fact that the signal effectively uses frequency diversity, in that all the transmitted information could be extracted from either the mark frequency or the space frequency (each regarded as OOK: on–off keying) alone. This is very beneficial for traffic on the HF band, where selective fading may cause one of the frequencies to fade out completely while the other is still usable. Using this characteristic to the full, it is possible to receive the data correctly when one tone is unavailable due to fading (using a 'slideback' detector), or even when it is being jammed by a strong continuous signal (using a 'Law assessor' [2]). Reliability of HF communications can be improved using an ARQ (automatic repeat request) system, such as that defined in Reference 3.

The need for higher signalling rates on long-haul routes using the HF band brought problems when using FSK. An HF signal received at a distance of several thousand kilometres may be received via several different paths, for which the spread of propagation time may be several milliseconds. Thus increasing the baud rate could result in the early path version of one symbol overlaying the late path version of the preceding one, resulting in ISI (intersymbol interference). One solution introduced by the UK Foreign and Commonwealth Office [4] used MFSK (multifrequency shift keying) at a 10-baud signalling rate. In each 100 ms symbol, it transmitted one of 32 different tones, each one representing an ITA2 character. Thus the character rate equalled the baud rate and the system provided a throughput equivalent to an FSK ITA2 system operating at 75 baud. In a later improvement [5], each character was transmitted as a sequence of two tones at a 20-baud rate. The tones were selected from a group of 6 (or 12) giving operation equivalent to ITA2 at 75 baud (or ITA5 at 110 baud).

FSK/FEK are early forms of digital modulation and although simple to implement and robust, they are not bandwidth-efficient, the OBW being many times the useful bit-rate. Other more efficient modulation methods have been developed, e.g. phase shift keying (widely used at VHF where propagation characteristics are rather more stable than at HF) and combined phase-and-amplitude keying (used in terrestrial microwave telephony links where conditions are usually very stable). In FSK there is no ambiguity as to whether a given tone represents a mark or a space, since one is higher in frequency than the other. However, in phase shift keying, the only thing that changes is the phase of the single RF carrier. At the receiver there is no way of knowing the transmitted phase. Even if the transmitter and receiver each had an ideal clock, the number of wavelengths in the over-the-air path is unknown. Consequently, PSK (phase shift keying) systems always use

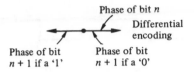

At transmitter

Phase of bit n

Differential encoding

Phase of bit $n + 1$ if a '1'

Phase of bit $n + 1$ if a '0'

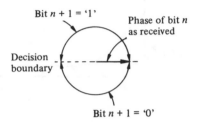

Phase of bit $n + 1$ if a '1'

Phase of bit n

Phase of bit $n + 1$ if a '0'

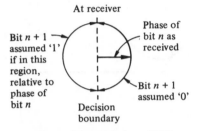

At receiver

Bit $n + 1$ assumed '1' if in this region, relative to phase of bit n

Phase of bit n as received

Bit $n + 1$ assumed '0'

Decision boundary

Differential demodulation of PSK

Bit $n + 1 = $ '1'

Phase of bit n as received

Decision boundary

Bit $n + 1 = $ '0'

(i) Asymmetrical PSK

(ii) Symmetrical form of PSK

a

Decision boundaries

01

AB
00

11

10

```
A  1
   0
B  1
   0
```
Same bit timing clock

QPSK carries 2 bits per symbol. Note Gray coding, so an error (phase on wrong side of boundary) will only affect A or B, not both.

(Asymmetrical form shown)

(i) QPSK

01   00

Bit n phase

11   10

SQPEK. In this symmetrical four-level system, the path taken between the vector at bit n and that at bit $n + 1$ (i.e. somewhere in one of the hatched areas), depends upon the preceding message bits.

(ii) SQPEK

b

Figure 7.4 *Various digital data modulation methods.*
a BPSK.
b Quadrature modulation (four-level, 2 bits/symbol). In (i), if the A data clock is offset by the half-bit period from the B data clock, the result is OQPSK, which has no 180° transitions.

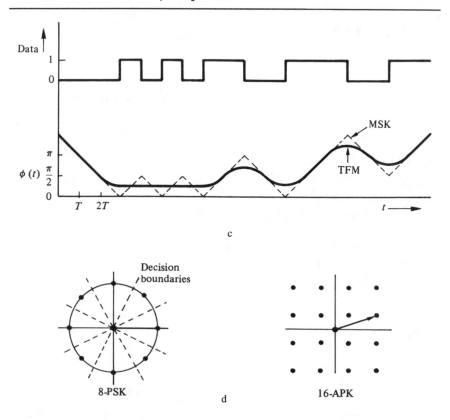

Figure 7.4 Continued
c *Tamed frequency modulation.*
d *Eight- and sixteen-level systems (3 and 4 bits/symbol, respectively).*

differential encoding (decoding may be either differential, or absolute, i.e. synchronous). Differential encoding means that a phase change from one symbol to the next indicates a one, and no phase change indicates a zero, or vice versa, depending upon the particular system. A transmission conse-quently needs a preamble of some sort, e.g. a series of ones, and this serves two purposes. Firstly, it enables the receiver to acquire symbol sync and secondly, the first zero following the ones can signal the start of the transmitted message. The simplest form of phase shift keying is BPSK (binary phase shift keying), often simply called PSK (see Figure 7.4a). The symmetrical form has the advantage that there is always a phase change so symbol sync (the same as bit sync for a binary modulation system) can always be maintained; in the unsymmetrical form a long string of zeros would result in no phase changes, so that the receiver's bit sync could drift out of synchronism. However, in the symmetrical form, a noise-induced phase shift at the receiver of only 90° (or less with differential decoding) will cause an error, whereas twice as large a phase shift is needed to give an error in the unsymmetrical form. Therefore, twice the received signal to noise ratio is necessary to prevent a noise-induced error, or put another way, half of the transmitted power is effectively dedicated to maintaining bit sync. On account of the 3 dB power advantage, unsymmetrical forms of PSK may be preferred (depending on the application), the modulation usually being of such a nature that long sequences of zeros do not occur. The receiver decides whether the phase of the signal during one bit is the same as or opposite to that in the preceding bit. The phase is sampled in the middle of the bit period, which is known from the bit-sync extraction circuit. Up to 90° difference counts as the same phase, more than this as the opposite phase. In differential decoding (DPSK), the bit phase is measured relative to the phase of the preceding bit, which may of course itself differ from the true phase due to noise. A further 3 dB reduction in the signal to noise ratio required for a given error rate is obtained if the measurement is made relative to true phase, i.e. synchronous decoding. This is possible if the phase of the original carrier is extracted, by doubling the frequency of the IF signal. Phase changes of 180° thus become 360° changes and an oscillator can then be phase locked to this signal. If the time constant of the phase lock loop filter is many times the bit period, the phase of the carrier is accurately recovered with minimal jitter, due to the averaging process.

Ideally, the OBW of the transmitted signal would be limited to $\pm f_b/2$ about the nominal carrier frequency, where f_b is the bit rate. However, if the phase changes in BPSK are instantaneous, there will be higher order sidebands (sidelobes), the first sidelobes being only 13 dB down. Filtering may be used to reduce the amplitude of these, but will have the effect of introducing amplitude variations into the envelope of the signal, which creates difficulties if the transmitter uses a class C power amplifier. It will also introduce ISI, resulting in a finite irreducible error rate on reception, even in the absence of noise. The ISI introduced by filtering can be largely corrected by a suitable all-pass filter or phase equalizer, but the problem of envelope variations remains. It can be minimized in some forms of QPSK (quadrature phase shift keying), also known as 4-level PSK. Here, there are four possibilities for each phase change, so each symbol conveys two bits of information (Figure 7.4b). The UK developed NICAM-728 (Near-Instantaneously Companded Audio Multiplex, providing digital-audio quality stereo or dual-language mono sound, adopted by the European Broadcasting Union for PAL and SECAM systems) uses asymmetrical QPSK. In other QPSK applications, the symmetrical form may sometimes be preferred, since then there is always an obvious minimum phase change to get from one symbol to another. In the unfiltered asymmetrical form, as in unfiltered asymmetrical BPSK, instantaneous 180° phase changes occur.

Instead of filtering, the phase transition can be arranged to occur smoothly, occupying an appreciable fraction of a symbol period, giving a much faster fall-off in sidelobe level without introducing envelope variations. SQPEK (four-level symmetrical differential phase exchange keying, Figure 7.4b) is produced by baseband filtering and pre-equalizing the data fed to I and Q (in-phase and quadrature) modulators and combining their IF outputs. It is a non-constant envelope scheme, exhibiting occasional dips in the envelope of up to 10 dB, depending upon the preceding bit sequence. To minimize both OBW and the receiver noise bandwidth, the overall filtering is equally split between transmitter and receiver. In the receiver IF the signal may be hard limited, but only after filtering to final bandwidth, otherwise excessive ISI is re-introduced. Bit rates up to 2400 bits/s are possible over HF paths using parallel tone modems. Reference 6 describes one such system, where 16 data tones and two special-purpose tones are transmitted continuously. Each data tone is BPSK or QPSK modulated at a 75 baud rate giving up to 2400 bits/s throughput in good conditions, with fall-back using increasing levels of diversity via 1200, 600 bits/s, etc., right down to 75 bits/s at 32 level diversity. However, with this scheme, the power available to each tone is very limited. Interest has therefore turned to serial tone modems for HF use, operating typically at 2400 bits/s. These use sophisticated filtering and training techniques to overcome the effect of ISI experienced due to the high baud rate, which is typically in excess of the effective bit rate to allow for periodic filter-training sequences, checkcodes, etc. Various formats are used, Reference 7 being one.

OQPSK (offset keyed QPSK, also known as OK-QPSK) and MSK (minimum shift keying, also known as FFSK and fast FSK) are important variants where the bit timing in the I and Q channels is offset by half a symbol period [8]. If either is band limited in the exciter to narrow the OBW and then hard-limited for the benefit of a class C power amplifier, the degree of regeneration of the filtered sidelobes is less than with filtered QPSK. Furthermore, MSK can be economically non-coherently detected using a discriminator, although a rather higher signal to noise ratio is then required. In unfiltered OQPSK (the asymmetrical form is usual), the maximum instantaneous phase change is 90°, since the component 180° I and Q channel phase changes are staggered. MSK and OQPSK may be coherently demodulated using the recovered carrier. This is obtained by quadrupling the IF signal, phase locking an oscillator to this and dividing its output by four. In MSK, as in CPFSK, there are no instantaneous phase transitions, so it offers low side sidelobe levels without the need for filtering, combined with a constant envelope. MSK can be viewed either as FSK where the frequency shift is $\pm 1/(4T)$, T being the bit period, or as OQPSK where the pulses in the I and Q modulator channels are shaped to a half-sinusoid instead of square. For a continuous stream of ones (or zeros), the phase of MSK advances (retards) linearly by 90° per bit period: for reversals (alternate 0s and 1s), it describes a triangular waveform of 90° peak-to-peak phase deviation. QMSK (quaternary MSK) is the symmetrical version, with phase changes of 45° or 135°: GMSK (Gaussian-filtered MSK) offers reduced sidelobe levels and these are even lower in QGMSK, which has been proposed for land mobile secure voice communications systems.

TFM (tamed frequency modulation) is a PR (partial response) version of MSK, offering even lower sidelobe levels at offsets from the carrier equal to the bit rate and beyond [9]. In a PR system, decoding one bit demands a knowledge of some other bits. In TFM, the bit information is spread over three adjacent bits, so that, for example, during a sequence of reversals the phase neither advances nor retards (Figure 7.4c). PR systems exhibit error propagation: an error in one bit may affect others also.

Where it is necessary to transmit a higher data rate in a given bandwidth than can be achieved with 4-level modulation, 8-PSK permits the transmission of three bits per symbol (Figure 7.4d) at the expense of requiring a higher E_b/N_0 (energy per bit over noise per unit bandwidth). Similarly, 16-PSK carries four bits per symbol, but as the number of levels increases, phase space positions become very crowded. Over high signal-to-noise ratio links, e.g. terrestrial microwave telephony bearers, the number of bits per symbol can be increased without such crowding by using both phase and amplitude modulation. Figure 7.4d shows 16-ary APK (sixteen level amplitude and phase keying); 64-ary APK is used on some links, carrying 6 bits per symbol.

For each type of modulation an appropriate demodulator is required in the receiver. Figure 7.5a shows a simple diode detector circuit for AM signals. The diode charges the RF bypass capacitor up to the peak voltage of the IF signal. A path to ground (or $-V_s$) is necessary to enable the voltage to fall again as the RF level falls on negative-going slopes of the modulating waveform. The detector circuit provides the demodulated audio frequency baseband signal varying about a dc level proportional to the strength of the carrier of the received signal. A capacitor blocks the dc level, passing only the audio to the volume control. The dc component across the RF bypass capacitor is extracted by a low-pass CR filter with typically a 100 ms time constant, and used as an AGC (automatic gain control) voltage to control the gain of the IF stages. This automatically compensates for variations of signal strength due to fading, and also ensures that weak and strong stations are all (apparently to the user) received at the same strength. Figure 7.5b shows one of the many forms of detector used for FM signals. A small winding closely-coupled to the primary of the discriminator transformer injects a signal V_{ref}, in phase with the primary voltage, at the centre tap of the secondary circuit, which is also tuned to 10.7 MHz. The secondary is very loosely magnetically coupled to the primary, so that the voltages V_1 and V_2 are in quadrature to the reference voltage when the frequency is exactly 10.7 MHz. As the frequency deviates about 10.7 MHz, V_1 and V_2 advance or retard (shown dotted) relative to V_{ref}, so the voltages VR_1 and VR_2 applied to the diodes become unequal, but R_1 and R_2 ensure that the average of VR_1 and VR_2 is held at ground potential. Thus the recovered audio appears at point A — note that the capacitor to ground at A is a short circuit to IF but an open circuit at audio frequency. (This circuit, known as the ratio detector, was popular in valve receivers in the early days of FM broadcasting as it provides a considerable degree of AM suppression. Thus if the level of the IF signal were suddenly to rise and fall (e.g. due to reflections from a passing vehicle or plane), the damping imposed upon the secondary would rise and fall in sympathy as the make-up current required to keep C_A charged to a higher or lower level varied. Modern FM receivers incorporate so much gain in the IF strip that they always operate with a hard-limited signal into the FM demodulator.) The recovered audio is de-emphasized to provide the mono-compatible sum signal; the stereo decoder extracts the difference signal from the raw recovered audio at point A. Figure 7.5c shows an FM quadrature detector. Here again the signal across the tuned circuit is in quadrature with the drive voltage when the frequency is exactly 10.7 MHz and varies in phase about this in sympathy with the deviation.

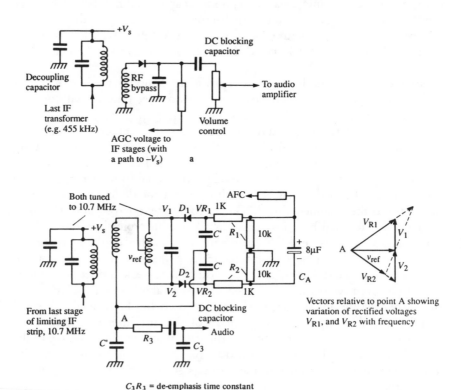

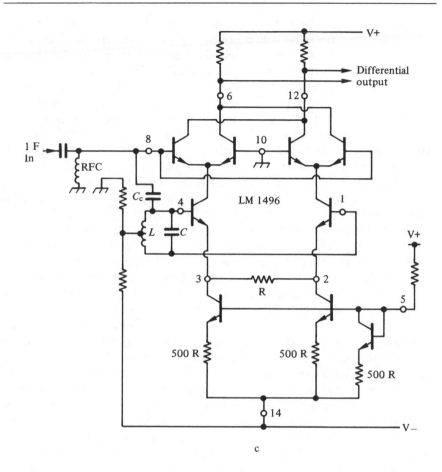

Figure 7.5 *AM and FM demodulators (detectors).*

a Diode AM detector. In the 'infinite impedance detector', a transistor base/emitter junction is used in place of the diode. The emitter is bypassed to RF but not to audio, the audio signal being taken from the emitter. Since only a small RF base current is drawn, the arrangement imposes much less damping on the previous stage, e.g. the last IF transformer, whilst the transistor, acting as an emitter follower, provides a low-impedance audio output.

b Ratio detector for FM, with de-emphasis. C' = RF bypass capacitor, 330 pF.

c Quadrature FM detector. Tuned circuit LC resonates at the Intermediate Frequency. C_c is small, so the signal at pins 1 and 4 is in quadrature with the IF input. R sets sensitivity (in volts per kilohertz deviation). Pin numbers refer to DIP (dual-in-line plastic) version of LM1496.

The phase detector output voltage thus varies about a steady dc level, in sympathy with the modulation. Both the ratio and the quadrature FM detectors provide a dc output level which is proportional to the standing frequency offset of the IF signal from 10.7 MHz. This voltage is usually fed back to control a varicap diode in the receiver's local oscillator circuit, in such a sense as to move the IF towards 10.7 MHz. This arrangement forms an AFC (automatic frequency control) loop, and if the loop gain is high, any residual mistuning is minimal. With the AFC in operation, as the receiver is slowly tuned across the band, it will snap onto a strong station and hold onto it until the receiver is tuned so far past it that the AFC range is exceeded, when it jumps out to the currently tuned frequency. It may thus be impossible to tune in a weak station on the adjacent channel to a strong one, so a switch is usually provided permitting the user to disable the AFC if required. Detectors for QAM and other signals using both phase and amplitude modulation are designed to be sensitive to both amplitude and phase variations. They also incorporate symbol timing extraction circuitry to determine exactly when in each symbol period to sample the signal. If operating as coherent detectors, they also need a carrier regeneration circuit.

Spread spectrum (SS) is a term indicating any of several modes of modulation which may be used for special purposes. Conceptually, the simplest form of SS is FH (frequency hopping), where the transmit frequency is changed frequently, usually many times per second. The transmit frequencies are selected in a pseudo-random sequence either from a predefined set of frequencies or from a block of adjacent channels. There is a dead time between each short tranmission or hop, typically of 10% of the hop dwell time, to allow the power to be ramped down and up again smoothly (avoiding spillage of spectral energy into adjacent channels) and to allow time for the synthesizer to change frequency. To minimize dead time, two synthesizers may be used alternately, allowing each a complete hop period to settle to its next frequency. The main purpose of an FH system is to provide security of the link against eavesdropping and exploitation, typically in an 'all-informed net' structure for tactical communications. Every station in the net will know the set of frequencies to be used and the PRBS (pseudo-random sequence); they also have pre-synchronized clocks driven from accurate frequency references, giving them a guide to the phase of the PRBS to within a few bit periods at worst. Periodic transmission of timing signals enables a late entrant to acquire net timing. By contrast, an adversary trying to penetrate the net does not know the set of frequencies in use and does not know the PRBS (which may be changed frequently for further security), let alone its phase. An FH system typically uses digital modulation, even though the traffic may be speech, which will be digitized and probably also encrypted. The bit rate over the air will be a little faster than the voice digitization rate, to allow for the dead periods; a FIFO (first in – first out memory) at the receiver reconstituting the original data rate. In order to receive the data transmitted during any one hop, the received signal to noise ratio in that particular channel must be at least as good as in a non-hopping link. Interference or jamming may wipe out any particular hop, but speech contains so much redundancy that up to 10% blocked channels is no disaster, especially at VHF where a higher hopping rate of several hundred per second (compared to nearer 10 hops/s at HF) can be used. Even jamming an FH system poses problems for an adversary; not knowing the exact channels in use, let alone their sequence, he must spread his available jamming power over the whole band. It will thus be much less effective than if he had been able to concentrate it on a single channel transmission.

The other type of SS is DS (direct sequence) spreading. This is used at VHF and UHF and is more versatile than FH. Whereas FH uses only one channel at a time, SS uses the whole band the whole of the time. This is achieved by deliberately increasing the bit rate and hence the bandwidth of the transmitted data. For example, the baseband bandwidth of a 100 kb/s data stream is 50 kHz, giving a minimum bandwidth needed for the PSK modulated transmission of 100 kHz. However, if each successive data symbol (bit) is exclusive ORed with a 10 Mb/s PRBS prior to PSK modulation, the transmitted bandwidth will now be 10 MHz. The PRBS does not repeat exactly each symbol; each symbol is multiplied by the next 100 bits of a very long PRBS. The PRBS is called the 'chipping sequence' and in the example given there are 100 chips per symbol. In the receiver, the signal is multiplied by the same PRBS in the correct phase, e.g. at IF using a double balanced mixer or a SAW convolver. This has the effect of de-spreading the energy and concentrating it all back into the original bandwidth. The received signal strength is thus increased by the amount of the 'processing gain', which in the example given is ×100 or 20 dB. By contrast, any interference such as a large CW or narrow band signal is spread out by the chipping sequence. Thus the signal can be successfully received even though the RF signal at the antenna is many decibels below noise and interference. The receiver in a DS spreading system has to acquire both symbol and bit (chip) sync in order to recover the transmitted data, by means much as described above for an FH system. Eavesdropping is even more difficult, since an adversary will not even know that a transmission is taking place if the signal in space is below noise.

References

1. Carson, J. R. Notes on the theory of modulation. *Proc. I.R.E.*, **10**, 57 (Feb. 1922)

2. Allnat, Jones and Law, Frequency diversity in the reception of selectively fading binary frequency-modulated signals. *Proc. I.E.E.*, **104 B**(14) pp. 98–100 (March 1957)
3. CCIR Recommendation 476–3 ITU, Geneva
4. Robin, Bayley, Murray and Ralphs. Multitone signalling system employing quenched resonators for use on noisy radio-teleprinter circuits. *Proc. I.E.E.*, **110**(9), pp. 1554–68 (September 1963)
5. Ralphs. An Improved 'Piccolo' MFSK modem for h.f. telegraphy. *The Radio and Electronic Engineer*, **52**(7) 321–330 (July 1982)
6. MIL-STD-188C section 7.3.5
7. NATO STANAG 4285 (Restricted)
8. Gronemeyer, S. and McBride, A. MSK and offset QPSK. *I.E.E.E. Trans. on Communications*, **Com-24**(8), pp. 809–20 (August 1976)
9. de Jager and Dekker. Tamed frequency modulation, a novel method to achieve spectrum economy in digital transmission. *I.E.E.E. Trans. Communications*, **Com-26**, pp. 534–42 (1978)

8
Oscillators

RF oscillators are used to produce the carrier wave which is required for a radio communications system. In the earliest days of 'wireless communication', spark transmitters were used; these produced bursts of incoherent RF energy containing a broad band of frequencies, although tuned circuits were soon introduced to narrow the band. However, valves and later transistors and FETs enable a single frequency oscillator to be produced. Typically, a tuned circuit is connected to the input of an amplifier, the output of which is coupled back into the tuned circuit. If it be arranged that at the resonant frequency of the tuned circuit, the gain from the input of the active device to its output, through the tuned circuit and back to its input again exceeds unity, then the inevitable small level of input noise of the active device will be amplified and will build up to a large continuous oscillation. The original noise will have been broadband, but the selectivity of the tuned circuit ensures that only the initial noise at the resonant frequency is amplified. Some mechanism is necessary to limit the amplitude of the oscillation and if one is not deliberately designed in then the circuit itself will provide it, for clearly the amplitude cannot go on building up for ever. Thus we have an oscillator with a steady output level at the frequency of the tuned circuit, plus the broadband noise of the device. The latter will still of course be there, though its level may be modified by the effect of the oscillator's amplitude determining mechanism reducing the amplifier's gain. The steady wanted output signal will in practice have very minor random amplitude and phase variations. The actual output can be resolved into an ideal output free of any amplitude or phase variations, plus random AM and PM noise sidebands: these fall off rapidly in amplitude with increasing offset from the wanted output frequency (Figure 8.1). The noise sidebands result in us being unable to predict at any instant exactly where in a 'circle of confusion' (much exaggerated in Figure 8.1) the tip of the vector is. The circle has no hard and fast boundary, the amplitude distribution with time of both the AM and FM noise sidebands exhibiting a normal or Gaussian distribution. In principle,

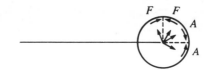

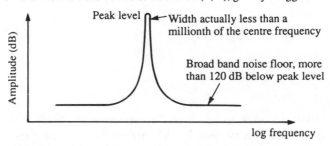

Sine wave with AM and FM noise sidebands (A, F), grossly exaggerated

Peak level → Width actually less than a millionth of the centre frequency

Broad band noise floor, more than 120 dB below peak level

Amplitude (dB)

log frequency

Corresponding frequency domain representation

Figure 8.1   *Real-life sinewave.*

the AM sidebands can be stripped off by passing the signal through a hard limiter, but any signal is necessarily accompanied by noise at thermal level or above and with a well-designed oscillator circuit, subsequent limiting will produce no significant reduction in AM noise sidebands. In any case, in most applications the PM noise sidebands are the most significant, as the most bandwidth-efficient modulation schemes (such as 8-ary PSK and others) are usually variants of phase modulation. The precise way in which the level of the PM sidebands drops off at increasing offsets from the carrier frequency depends upon a number of factors [1], but before considering this, note that an oscillator will also exhibit long-term frequency variations and these are best considered in the time domain.

Consider an oscillator circuit which is running continuously for a long period. Over a time scale of days to years there will be a gradual drift in the oscillator's frequency, due to ageing of the components. For example, in an *LC* oscillator, it is difficult to produce an inductance with a long-term stability better than 1 part in 10^4. Where this is inadequate, a crystal oscillator may be used. The resonant frequency of a crystal will also drift with time. In the case of a solder-seal metal-can crystal the drift will usually be negative (falling frequency) due to the very small but finite vapour pressure of lead resulting in the deposition of lead atoms on the crystal. With cold-weld and glass-encapsulated types the drift is considerably less and may be either positive or negative. In the medium term, minutes to days, an oscillator will also exhibit frequency variations with changes in temperature due to the tempcos of the various components; here again crystal oscillators outperform LC types.

Returning to short-term variations, over periods of a few seconds or less, these are usually considered in the frequency domain as $\mathcal{L}(f_m)$ dBC, the ratio of the single-sided phase noise power in a 1 Hz bandwidth to the carrier power (expressed in decibels), as a function of the offset-frequency (also called sideband-, modulation- or baseband-frequency) from the carrier. In practice, this is measured with a spectrum analyser, the result being the same whether the offset from the carrier at which the measurement is made be positive or negative, since the noise spectrum is symmetrical about the carrier (Figure 8.1). The following regions may be distinguished, moving progressively away from the carrier. At a very small offset fHz the power is proportional to f^{-4}, i.e. a 12 dB/octave roll-off (the random walk FM region); as f increases this changes to f^{-3} (-9 dB/octave, flicker FM), then f^{-2} (-6 dB/octave, random walk phase), then f^{-1} (-3 dB/octave, flicker phase). The latter continues until the f^0 region of flat far-out noise floor is reached: this cannot be less than -174 dBm (thermal in a 1 Hz bandwidth) and is typically -150 dBC or better. The breakpoints between the regions are gradual and where two are fairly close together, the corresponding region may not be observed at all. More details can be found in Reference 2.

Turning to practical oscillators, Figure 8.2b shows a schematic filter/amplifier type oscillator, as described at the beginning of the chapter. Figure 8.2a shows a negative resistance type oscillator, examples being the Hartley and Colpitts circuits. In this type of oscillator, an active device is connected across a tuned circuit in such a way as to reflect a negative resistance $-R_d$ in parallel with the tuned circuit, where R_d is the dynamic resistance of the tuned circuit. Thus the net losses are just made up, raising the effective Q to infinity at that particular level of oscillation. At lower levels, the negative resistance reflected across the tuned circuit is numerically lower, resulting in a loop gain exceeding unity, whilst at higher levels the negative resistance would be numerically greater than R_d, resulting in the losses in the tuned circuit exceeding the energy supplied by the active device. In practice, there is no real difference between the negative resistance and the filter/amplifier views of an oscillator and the latter is convenient for purposes of explanation. Figure 8.3 shows plots of loop gain from the input of the amplifier to its output, through the filter (tuned circuit) and back again to the input, versus the input signal level to the amplifier. Characteristic 8.3c is typical of a well-designed oscillator: the loop gain at low levels exceeds unity by a comfortable margin and passes through unity at a steep angle. Such an oscillator is a sure-fire starter and the output level is very stable with low AM noise sidebands. Characteristic 8.3a is also met and is often acceptable, but 8.3b represents a totally unsatisfactory design. Such an oscillator will often start despite the less than unity small signal gain, due to the switch-on transient, but may fail to operate occasionally. Characteristic 8.3d represents

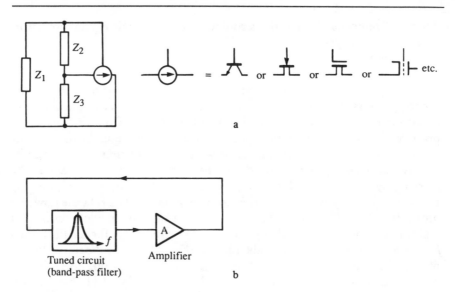

Figure 8.2 *Oscillator types.*
a Negative resistance oscillator: see text.
b Filter/amplifier oscillator.

an oscillator specially designed so that its gain changes only very gradually with level. Its amplitude of oscillation is consequently very susceptible to outside influences and such a circuit (coupled to a detector) will receive SW broadcast and amateur transmissions without an aerial of any sort connected when the loop gain is adjusted so that oscillation just commences, operating as a synchrodyne receiver.

The negative resistance oscillator of Figure 8.2a will only oscillate if Z_2 and Z_3 are reactances of the same sign and Z_1 is of the opposite sign. Z_1 capacitive gives the Hartley family of oscillators and Z_1 inductive gives the Colpitts and its derivatives, the Clapp and Pierce oscillators. These are shown in Figure 8.4 along with sundry other types, including the TATG (tuned anode, tuned grid), so called from its valve origins. In the Clapp oscillator, noted for its good frequency stability, the additional capacitor C_1 acts, together with C_2 and C_3, as a step-down transformer. This reduces the shunting effect on the tuned circuit of the input and output conductances and susceptances of the active device. Due to the light coupling of the active device to the tuned circuit, the arrangement requires an active device with a high mutual conductance, giving a large power gain. The dual-gate MOS-FET electron-coupled oscillator is the solid state equivalent of the grounded screen valve tetrode circuit. (There is no solid state equivalent of the grounded cathode electron coupled oscillator, since that needs a pentode.)

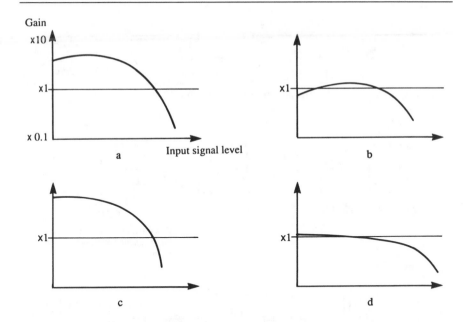

Figure 8.3 *Oscillator feedback: degree of coupling.*
a–d Characteristics (see text).

The electron-coupled circuit acts as both oscillator and buffer stage, variations of loading on the drain circuit having very little effect on the frequency.

Figure 8.5 shows filter/amplifier oscillators of various sorts. The line-stabilized oscillator (like the line-stabilized TATG) is restricted to UHF and above, where a line of length equal to half a wavelength or more becomes a manageable proposition. At UHF, SAW delay lines can provide a delay of many cycles with little insertion loss and good stability. There is thus a 'comb' of frequencies at which they exhibit zero phase shift. A tuned circuit is required to select the desired frequency of oscillation: if the capacitor is a varactor, then one of a number of possible frequencies can be selected as required. Figure 8.6 shows oscillator circuits using two active devices. The greater maintaining-circuit power-gain available in the Franklin oscillator permits lighter coupling to the tuned circuit, reducing the pulling effect of stray maintaining circuit reactances. On the other hand, the additional device means that there is now another source of possible phase-shift variations round the loop. The emitter-coupled circuit of Figure 8.6b is unusual in that the tuned circuit operates at series resonance. It is thus suitable for a crystal operating at or near series resonance. This generally provides greater frequency stability than operation at parallel resonance,

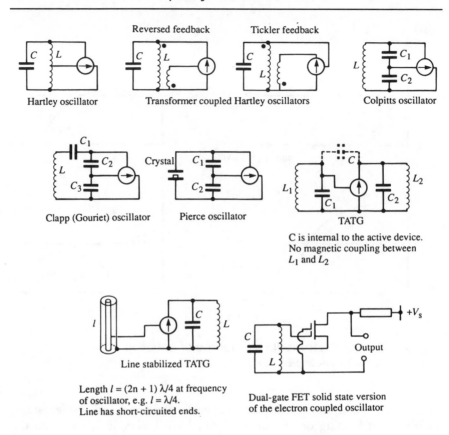

Figure 8.4 *Negative resistance oscillators (biasing arrangements not shown).*

although the available pulling range is only about a tenth of that of a parallel-resonant crystal oscillator such as in Figure 8.4.

Figure 8.7a shows another oscillator circuit using two active devices, this time in push–pull. The two devices operate in antiphase but are effectively in parallel; it is not an emitter-coupled circuit. This arrangement elegantly solves one of the problems encountered with a single device bipolar transistor oscillator such as in Figure 8.4. In those circuits, the amplitude of oscillation usually increases until the net gain is brought down to unity by collector saturation imposing heavy damping on the tuned circuit at the negative peaks of collector voltage excursion (assuming an NPN implementation). It is usual to arrange that the resultant increase in base current biases the transistor back to a lower average collector current where the gain is also lower, but the increased damping is an undesirable (and usually the major) effect which stabilizes the amplitude. This effect did not arise in valve

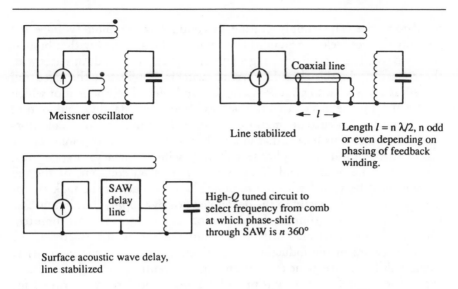

Meissner oscillator

Line stabilized

Length $l = n\ \lambda/2$, n odd or even depending on phasing of feedback winding.

High-Q tuned circuit to select frequency from comb at which phase-shift through SAW is $n\ 360°$

Surface acoustic wave delay, line stabilized

Figure 8.5 *Filter/amplifier oscillators.*

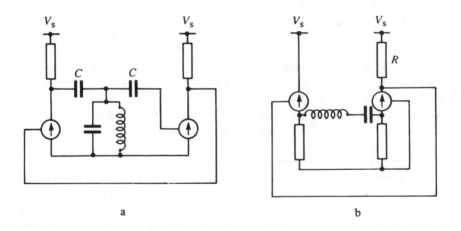

a

b

Figure 8.6 *Two-device oscillators.*

a Franklin oscillator. The two stages provide a very high non-inverting gain. Consequently the two capacitors C can be very small and the tuned circuit operates at close to its unloaded value of Q.

b Butler oscillator. This circuit is unusual in employing a series tuned resonant circuit. Alternatively it is suitable for a crystal operating at or near series resonance, in which case R can be replaced by a tuned circuit to ensure operation at the fundamental or desired harmonic, as appropriate.

oscillators, the valve simply ceasing to conduct as the anode voltage fell towards or even below ground. (The same can be arranged with a bipolar transistor oscillator by connecting a high speed Schottky diode in series with the collector.) In the class D current switching oscillator, the fixed tail current is chopped into a squarewave, the fundamental component of which is selected by the tank circuit. For best frequency stability and output waveform, the tail current should be set at such a value that the transistors do not bottom. This means that in a wide range oscillator, one must either accept that the output amplitude will vary with frequency, or one must arrange to tune both L and C so as to maintain R_d constant, or the tail current must be varied with the tuning. The centre tap of the tank circuit may be connected directly to the decoupled positive supply, but in this case the centre tap to ground of the tuning capacitance is best omitted. Otherwise, problems may arise if the inductor tap is not exactly at the electrical centre of the inductor — effectively giving two tuned circuits at slightly different frequencies. Grounding the centre point of the tuning capacitance is preferred since it provides a near short circuit to ground for the unwanted harmonic components of the device collector currents. These will be considerable, assuming the two resistors R are set to zero, as will usually be the case; the resistors may be added if desired to produce a characteristic approaching that in Figure 8.3d. If one of the two cross-coupling capacitors C is omitted, the circuit operates as an emitter-coupled negative-resistance oscillator, preserving some of the better characteristics of the original.

Figure 8.7b and c show two clock oscillators such as are used in microprocesor systems. The first operates at the series resonant frequency of the crystal; capacitor C provides some phase advance to compensate for the lag due to the propagation delay of the inverters. The second operates with the crystal near parallel resonance; component values will depend upon the operating frequency. In cost-sensitive applications the crystal can often be replaced by a ceramic resonator. In applications where frequency stability is the prime consideration, such as the frequency reference for a synthesizer, the rough and ready crystal oscillators of Figure 8.7 would be replaced by a TCXO (temperature-compensated crystal oscillator) or an OCXO (oven-controlled crystal oscillator). In the latter, the crystal itself and its maintaining circuit are housed within a container, the interior of which is maintained at a constant temperature higher than the highest expected ambient temperature, commonly at $+75°C$. A OCXO can provide a tempco of output frequency in the range 10^{-7}–10^{-9} per °C, but stabilities substantially better than one part in 10^6 per annum are difficult to achieve with an AT cut crystal, although recent developments have improved on this to 1 in 10^9 per annum (typical), with phase noise already down to -140 dBc at only 10 Hz offset from the carrier. Figure 8.8a shows the typical cubic or 'S'-shaped frequency variation of an AT cut crystal with temperature. The AT cut is

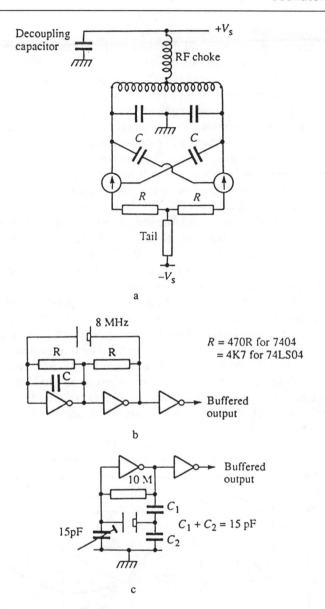

a

b

c

Figure 8.7

a *Class D or current switching oscillator; also known as the Vakar oscillator. With R zero, the active devices act as switches, passing push–pull square waves of current. Capacitors C may be replaced by a feedback winding. R may be zero, or raised until circuit only just oscillates. 'Tail' resistor approximates a constant current sink.*

b *TTL type with crystal operating at series resonance.*

c *CMOS type with crystal operating at parallel resonance.*

'singly rotated': one of the crystallographic axes lies along a diameter of the crystal blank but the orthogonal diameter of the blank is slightly offset from the orthogonal axis. By selecting the offset angle, the tempco at the point of inflection (which occurs at around 29°C) can be set anywhere from positive through zero to negative. It is thus possible in a non-temperature controlled oscillator to have a very low frequency variation with temperature over a rather limited range centre on 29°C, whilst if a larger temperature range must be covered then the angle of cut will be increased, leading to larger frequency variations with temperature. If an AT cut crystal is to be used in an OCXO, then again an increased angle of cut will be used, such as to place the upper turn-over point at the oven temperature (Figure 8.8b). The short- to medium-term stability of an OCXO is optimum when it is operated continuously. On the other hand, the long-term stability is then worse, since ageing is faster at oven temperature than at ambient. Figure 8.8b also shows the temperature variations of the BT and SC cuts in the region of the oven temperature. The SC (strain compensated) cut is doubly rotated, i.e. none of the three orthogonal crystallographic axes lies in the plane of the crystal blank. The SC cut is therefore more complicated to produce and hence more expensive than other types, but it offers improved resistance to shock and superior ageing performance. However, care in application is required, since the SC cut also exhibits more spurious resonance modes. For example, the 10 MHz SC crystal used in the Hewlett-Packard 10811A/B ovened reference oscillator is designed to run in the third overtone C mode resonance. The third overtone B mode resonance is at 10.9 MHz, the fundamental A mode resonance is at 7 MHz, and below that are the strong fundamental B and C

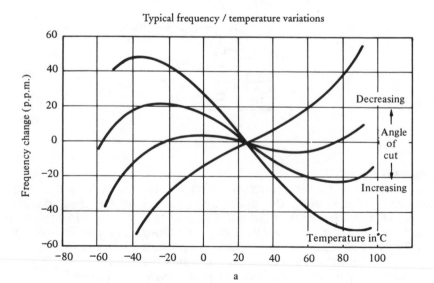

a

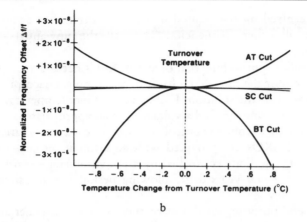

b

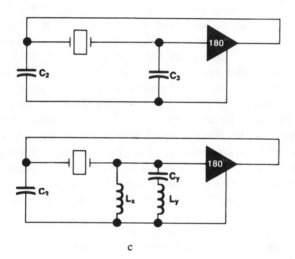

c

Figure 8.8

a Temperature characteristics of AT cut crystals (reproduced by courtesy of SEI Ltd, a GEC company).

b Temperature performance of SC, AT and BT crystal cuts.

c Standard Colpitts oscillator (top) and the same oscillator with SC mode suppression (10811A/B oscillator). (Reproduced with the permission of Hewlett-Packard Co.)

modes. Figure 8.8c shows the SC cut crystal connected in what is basically a Colpitts oscillator, so as to provide the 180° phase inversion at the input of the inverting maintaining amplifier. With the correct choice of L_x, L_y and C_y, they will appear as a capacitive reactance over a narrow band of frequencies centred on the desired mode at 10 MHz, but as an inductive reactance at all other frequencies. Thus all the unwanted modes are suppressed [3].

Where stability approaching that of an OCXO is necessary but the power drain of an oven or the time taken for it to warm up is unacceptable, then a TCXO may provide the solution. In this, the ambient temperature is sensed by one or more thermistors and a voltage with an appropriate law is derived for application to a voltage-controlled variable capacitor (varicap). Both OCXOs and TCXOs are provided with adjustment means — a trimmer capacitor or varicap diode controlled by a potentiometer — with sufficient range to cover several years drift, allowing periodic re-adjustment to the nominal frequency.

For a general-purpose signal source such as a signal generator for the laboratory or test department, the traditional solution was an LC oscillator with switch selection of several ranges, accurately calibrated. Often a 1 MHz or 10 MHz crystal oscillator was incorporated, so that one of its harmonics could be used to check the scale calibration at the nearest 1 or 10 MHz point. Later, some signal generators were provided with 'lock boxes'. Here, a variable ratio divider was set by the user to the appropriate setting for the RF output frequency of the signal generator, whose frequency was thus locked to that of the lock box's crystal reference via the generator's dc coupled external FM modulation input. In a still later development, the generator was equipped with a counter which both indicated the output frequency and provided the lock box setting, as in the legendary Hewlett-Packard 8640 series. When a LOCK button was pressed, a PLL (phase lock loop) was implemented as with the earlier separate lock boxes. It was not long before the operation of the PLL was entirely automated, making its operation transparent to the user. PLLs are now widely applied to frequency sources of all sorts in addition to signal generators, for example the local oscillators used in transmitters and receivers (see Chapter 10). Figure 8.9 shows the generic block diagram of a PLL and illustrates the operation of a first-order loop. A sample of the output of the VCO (voltage-controlled oscillator) is fed via a buffer amplifier to a variable ratio divider, e.g. ratio N. The divider output is compared with a comparison frequency f_c, derived by dividing the output of a stable reference frequency source f_{ref}, such as a crystal oscillator, by a fixed reference divider ratio M. An error voltage is derived which, after smoothing, is fed to the VCO in such a sense as to reduce the frequency difference between the variable ratio divider's output and the comparison frequency. If the comparison is performed by a frequency discriminator there will be a standing frequency error in the synthesizer's output, albeit small if the loop gain is high. Such an arrangement is called a frequency lock loop (FLL); these are used in some specialized applications. However, the typical modern synthesizer operates as a PLL, where there is only a standing phase difference between the ratio N divider's output and the comparison

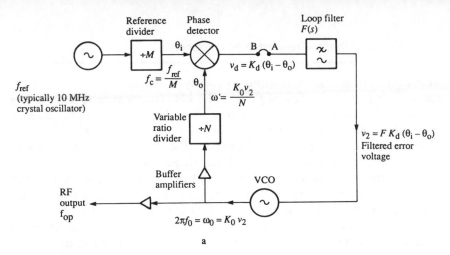

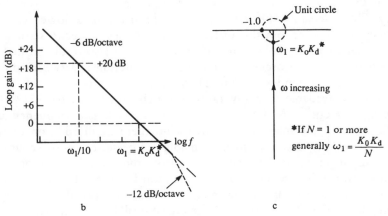

b

c

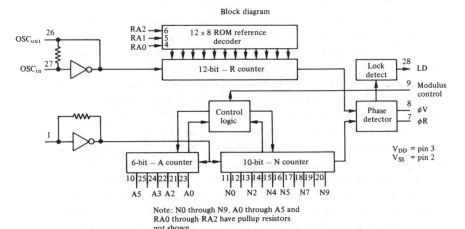

Figure 8.9

a Phase lock loop synthesizer.

b Bode plot, first0order loop.

c Nyquist diagram, first-order loop.

d Block diagram of an LSI variable ratio N divider, with a counter to control a two modulus P.P + 1 prescaler, Motorola type MC145152. (Reproduced by courtesy of Motorola Ltd.)

frequency. The oscillator's output frequency is simply Nf_c, where f_c is the comparison frequency. Thus if f_c were 12.5 kHz (Europe) or 15 kHz (USA) we would have a simple means of generating any of the transmit channel frequencies used in the VHF private mobile radio (PMR) band.

In fact there is a practical difficulty in that variable ratio divide-by-N counters which work up to VHF or UHF frequencies are not available, but this problem is circumvented by the use of a prescaler. If a fixed prescaler ratio, say divide by 10, were used, then in the PMR example, the comparison frequency would have to be reduced to 1.25 kHz to compensate. However, the lower the comparison frequency, the more difficult it is to avoid comparison frequency ripple at the output of the phase comparator passing through the loop filter and reaching the VCO, causing comparison frequency FM sidebands. Of course we could just use a lower cut-off frequency in the filter, but this makes the synthesizer slower to settle to a new channel frequency following a change in N and also results in higher noise sidebands in the oscillator's output. The solution is a two-modulus prescaler such as a divide by 10 or 11 type, usually written $\div 10/11$. Such prescalers are available in many ratios through $\div 64/65$ up to $\div 512/514$, providing a 'fractional N' facility so that a high comparison frequency can still be used. In the main loop divider chip there is, in addition to the programmable $\div N$ counter, a programmable $\div A$ prescaler-control counter. After A input pulses to the main divider from the prescaler, the former's prescaler control line switches the prescale ratio from $P + 1$ to P, where it remains until the main divider has received N pulses, when the prescaler is switched back to $\div(P + 1)$. If $A = 0$ then the overall divide ratio N_{total} from the prescaler plus main divider is simply $\div PN$. For any value of A, every pulse out of the main divider will require extra A pulses into the prescaler, so that $N_{total} = PN + A$. Thus if $A = N/2$, then $N_{total} = N(P + \frac{1}{2})$, hence the term 'fractional ratio divider' for the combination of main and prescale counters. P can be incremented from P to $P + 1$ in steps of $1/N$, enabling $NP = N_{total}$ to be effectively incremented in steps of unity between values of the main divider N (see Figure 8.9d). Clearly, A must not be greater than N; also $N_{total;min} = (P - 1)P + A$ and $N_{total;max} = N_{max}P + A_{max}$. Other constraints will apply in any given situation, due to propagation times through the main and prescale counters and to the latter's set-up and release times relative to its modulus control input.

A PLL synthesizer is an NFB loop and, as with any NFB loop, care must be taken to roll off all the loop gain safely before the phase shift reaches 180°. This is easier if the loop gain does not vary wildly over the frequency range covered by the synthesizer. Hence a VCO whose output frequency is a linear function of the control voltage is an advantage. The other elements of the loop also need to be correctly proportioned and the parameters of these have been marked in Figure 8.9a, following for the most part the terminology used in what is probably the best known treatise on phase lock loops [4]. Assuming that the loop is in lock, then both inputs to the phase detector are at the comparison frequency f_c, but with a standing phase difference $\theta_i - \theta_o$. This results in a voltage v_d out of the phase detector equal to $K_d(\theta_i - \theta_o)$.

In fact, the phase detector output will usually include ripple at the comparison frequency or at $2f_c$, although there are phase detectors which produce very little (ideally zero) ripple. The ripple is suppressed by the low-pass loop filter, which passes v_2 (the dc component of v_d) to the VCO. Assuming that the VCO's output radian frequency ω_0 is linearly related to v_2, then $\omega_0 = K_0 v_2 = K_0 F K d(\theta_i - \theta_0)$, where F is the response of the low-pass filter. Because the loop is in lock, ω' (i.e. ω_0/N) is the same radian frequency as w_c, the comparison frequency. If the loop gain $K_0 F K_d/N$ is high, then for any frequency in the synthesizer's operating range, $\theta_i - \theta_0$ will be small. The loop gain must be at least high enough to tune the VCO over the frequency range without $\theta_i - \theta_0$ exceeding $\pm 90°$ or $\pm 180°$, whichever is the maximum range of the phase detector being used.

Let us check up on the dimensions of the various parameters. K_d is measured in volts per radian phase difference between the two phase detector inputs. F has units simply of volts per volt at any given frequency. K_0 is in hertz per volt, i.e. radians per second per volt. Thus whilst the filtered error voltage v_2 is proportional to the difference in phase between the two phase detector inputs, v_2 directly controls not the VCO's phase, but its frequency. Any change in frequency of ω_0/N, however small, away from exact equality with εc will result in the phase difference $\theta_i - \theta_o$ increasing indefinitely with time. Thus the phase detector acts as a perfect integrator, whose gain falls at 6 dB per octave from an infinitely large value at dc. It is this infinite gain of the phase detector, considered as a frequency comparator, which is responsible for there being zero net average frequency error between the comparison frequency and f_{op}/N. Consider a first order loop, i.e. one in which the filter F is omitted, or where $F = 1$ at all frequencies, which comes to the same thing. At some frequency ω_1 the loop gain, which is falling at 6 dB/octave due to the phase detector, will be unity (0 dB). This is illustrated in Figure 8.9b and c, which shows the critical unity loop gain frequency ω_1 on both an amplitude (Bode) plot and a vector (Nyquist) diagram. To find ω_1 in terms of the loop parameters K_0 and K_d without resort to the higher mathematics, we can notionally break the loop at B, the output of the phase detector, and insert at A a dc voltage exactly equal to that which was there previously. Now superimpose upon this dc level a sinusoidal signal, say a 1 V peak. The resultant peak FM deviation of ω_o will be K_0 rad/s. If the frequency of the superimposed sinusoidal signal were itself K_0 rad/s, then the modulation index would be unity, corresponding to a peak VCO phase deviation of ± 1 rad (see Chapter 7). This would result in a deviation of $\pm 1/N$ rad at the phase detector input and hence a detector output of K_d/N volts. If we change the frequency of the input at A from K_0

to $K_0 K_d/N$, the peak VCO phase deviation will now be N/K_d. The deviation at the phase detector input is thus $1/K_d$ and so the voltage at B will be unity. So the unity loop gain frequency ω_1 is $K_0 K_d/N$ rad/s, as shown in Figure 8.9b and c. With a first order loop there is no independent choice of gain and bandwidth, quite simply $\omega_1 = K_0 K_d/N$. We could re-introduce the filter F as a simple passive CR cutting off at a corner frequency well above ω_1, as indicated by the dotted line in Figure 8.9b and by the teacup handle at the origin in Figure 8.9c, to help suppress any comparison frequency ripple. This technically makes it a low-gain second order loop, but it still behaves basically as a first order loop provided the corner frequency of the filter is well clear of ω_1 as shown.

Synthesizers usually make use of a high-gain second order loop, which will be examined in a moment, but first a word as to why this type is preferred. Figure 8.10a compares the close in spectrum of a crystal oscillator with that of a mechanically-tuned LC oscillator and a VCO. Whereas the output of an ideal oscillator would consist of energy solely at the wanted output frequency f_0, that of a practical oscillator is accompanied by undesired noise sidebands, representing minute variations in the oscillator's amplitude and frequency. In a crystal oscillator these are very low, so the noise sidebands, at 100 Hz either side, are typically -120 dB relative to the wanted output, falling to a noise floor further out of about -150 dB. The Q of an LC tuned circuit is only about one hundredth of the Q of a crystal, so the noise of a well-designed LC oscillator reaches -120 dB at more like 10 kHz off tune. In principle, a VCO using a varicap should not be much worse than a conventional LC oscillator provided the varicap diode has a high Q over the reverse bias voltage range, but with the high value of K_0 commonly employed (maybe 10 MHz/V or more) noise on the control voltage line is a potential source of degradation. Like any NFB loop, a phaselock loop will reduce distortion in proportion to the loop gain. 'Distortion' in this context includes any phase deviation of ω', and hence of ω_0, from the phase of the comparison frequency. Thus over the range of offset from the carrier for which there is a high loop gain, the loop can clean up the VCO output to something more nearly resembling the performance of the reference, as illustrated in Figure 8.10b.

A second order loop enables us to maintain a high loop gain up to a higher frequency, by rolling off the loop gain faster. Consider the case where the loop filter is an integrator as in Figure 8.11c; this is an example of a high-gain second order loop. With the 90° phase lag of the active loop filter added to that of the phase detector, there is no phase margin whatever at the unity gain frequency; as Figure 8.11b shows, we are heading for disaster (or at least instability) at ω_1 where the loop gain is unity; $\omega_1 = F K_0 K_d/N$. By reducing the slope of the roll-off in Figure 8.11a to 6 dB/octave before the frequency reaches ω_1 (dotted line), we can restore a phase margin, as shown dotted in Figure 8.11b, and the loop is stable. This is achieved simply by

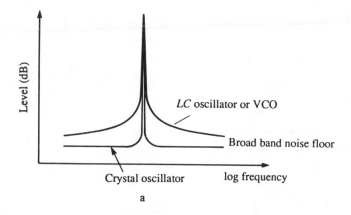

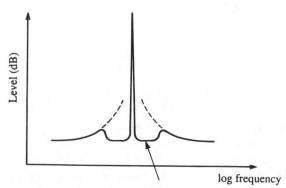

Figure 8.10 *Purity of radio-frequency signal sources.*
a Comparison of spectral purity of a crystal and an LC oscillator.
b At low-frequency offsets, where the loop gain is still high, the purity of the VCO (a buffered version of which forms the synthesizer's output) can approach that of the crystal derived reference frequency, at least for small values of N/M.

inserting a resistor R_2 in series with the integrator capacitor C at X–Y in Figure 8.11c. This is the active counterpart of a passive transitional lag. If we make $R_1 = \sqrt{2}.R_2$, then at the corner frequency of the filter $\omega_f = 1/(CR_2)$ the gain of the active filter is unity and its phase shift is 45°, whilst at higher frequencies it tends to −3 dB and zero phase shift. If we make ω_f equal to K_0K_d/N, then ω_1 (the loop unity gain frequency) is unaffected but there is now a 45° phase margin. It is convenient if K_0, K_d and N are dimensioned so

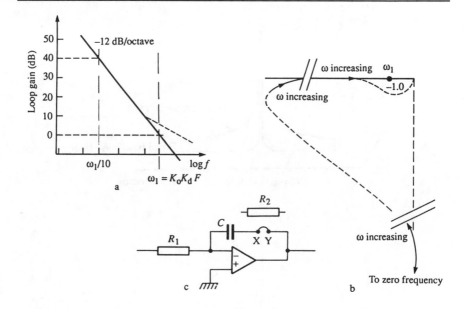

Figure 8.11 *PLL with second-order active loop filter (see text).*

that the corresponding first order loop unity-gain frequency $\omega_1 = K_0K_d/N$ is about one-tenth or less of the comparison frequency f_c. Otherwise it becomes more difficult to avoid phase comparator ripple causing comparison frequency FM sidebands on the VCO output. If necessary, a comparator frequency notch filter can be included in the loop.

As Figure 8.11a shows, at frequencies well below ω_1, the loop gain climbs at 12 dB/octave accompanied by a 180° phase shift, until the op-amp runs out of open loop gain. This occurs at the frequency ω where $1/(\omega C)$ equals A times R_1, where A is the open loop gain of the op-amp (an op-amp integrator only approximates a perfect integrator). Below that frequency, the loop gain continues to rise for evermore, but at just 6 dB/octave with an associated 90° lag, due to the phase detector which, as we noted, is a perfect integrator. This change occurs at a frequency too low to be shown in Figure 8.11a; it is off the page to the top left. It is only shown in Figure 8.11b by omitting chunks of the open-loop locus of the tip of the vector.

For a high-gain second order loop, analysis by the root locus method [5] shows that the damping (phase margin) increases with increasing loop gain, so provided that the loop is stable at that output frequency (usually the top end of the tuning range) where K_0 is smallest, then stability is assured. This is also clear from Figure 8.11. For if K_0 or K_d increases, then so will ω_1, the unity gain frequency of the corresponding first order loop. Thus ω_1 is now higher than ω_f (the corner frequency of the loop filter), so the phase margin will now be greater than 45°. Having found a generally suitable filter, let us return for another look at phase detectors and VCOs. Figure 8.12 shows several types of phase detector and indicates how they work. The logic types are fine for an application such as a synthesizer, but not so useful when trying to lock onto a noisy signal, e.g. from a distant, tumbling, space-craft — here the EXOR type is more suitable, in conjunction perhaps with a third order loop to give minimal frequency error with changing Doppler shift of the incoming signal. Both pump-up/pump-down and sample-and-hold types exhibit very little ripple when the standing phase error is very small, as is the case in a high-gain second order loop. However the pump-up/pump-down types can cause problems. Ideally, pump-up pulses — albeit very narrow — are produced however small the phase lead of the reference with respect to the variable ratio divider output; likewise pump-down pulses are produced for the reverse phase condition. In practice, there may be a very narrow band of relative phase shift around the exactly in-phase point, where neither pump-up nor pump-down pulses are pro-duced. The synthesizer is thus an entirely open loop until the phase drifts to one end or other of the 'dead space', when a correcting output is produced. Thus the loop acts as a 'bang-bang' servo, bouncing the phase back and forth from one end of the dead space to the other — evidenced by unwanted noise sidebands. Conversely, if both pump-up and pump-down pulses are pro-duced at the in-phase condition, the phase detector is no longer ripple-free when in lock and, moreover, the loop gain may rise at this point. Ideally, the phase detector gain K_d should, like the VCO gain K_0, be constant. Constant gain, and absence of ripple when in lock, are the main attractions of the sample-and-hold phase detector. In the quest for low-noise sidebands in the output of a synthesizer, many ploys have been adopted. One very powerful aid is to minimize the VCO noise due to noise on the tuning voltage, by substantially minimizing K_0, to the point where the error voltage can only tune the VCO over a fraction of the required frequency range. The VCO is pre-tuned by other means to approximately the right frequency, leaving the phaselock loop with only a fine tuning role. Figure 8.13 shows an example of this arrangement [6].

There are alternatives to the PLL approach to frequency generation. One of these is the direct synthesizer, pioneered by General Radio. A develop-ment of this system, using binary rather than decade increments in frequen-cy resolution, has been developed by Eaton Instruments (AILtech Division). In this scheme there is no effective frequency multiplication, as there is in a PLL. Instead, the required output frequency is built up by successively mixing selected harmonics of the very pure quartz crystal derived reference frequency, giving an output with levels of close-in noise not much worse than a crystal oscillator, and not approached by PLL type generators. Another approach is DDS, direct digital synthesis — not to be confused with direct synthesis. In a DDS, a frequency setting number (held in a

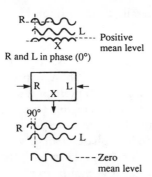

R and L in phase (0°)

90°

R and L in quadrature (90°)

a

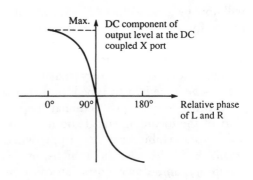

DC component of
output level at the DC
coupled X port

Relative phase
of L and R

A   0
B   0
C   0

$C = A \otimes B$

A   0
B   0
C   0

b

Max.   1

DC component
of C

0.5

0°   90°   180°

A   1
B   1
     0
     0

Pump-up
pulses

0

Pump-down
pulses

A
B

Pump-up output (PU)
Pump-down output (PD)

Maybe combined on
a single output pin

0
0
0
0
−1

PU

PD

c

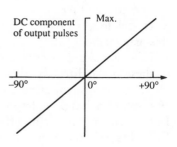

DC component
of output pulses

Max.

−90°   0°   +90°

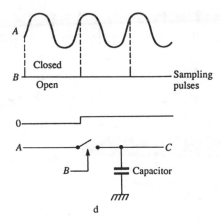

Figure 8.12   *Phase detectors used in phase lock loops (PLLs).*
a The ring DBM used as a phase detector is only approximately linear over say ±45° relative to quadrature.
b The exclusive-OR gate used as a phase detector.
c One type of logic phase detector.
d The sample-and-hold phase detector. In the steady state following a phase change, this detector produces no comparison frequency ripple.

register) is repeatedly added into an accumulator at each occurrence of a clock pulse. The top N bits of the accumulator (typically $N = 8$) are used to address a sine look-up ROM (read-only memory), the output values from which are passed to a DAC (digital to analog converter). Thus the latter outputs a stepwise approximation to a sinewave, each cycle corresponding to one pass through the ROM address range. An advanced implementation, using an arrangement needing just a quarter of a sinewave stored in ROM, is shown in Figure 8.14. At exceedingly low frequencies, the level correspond-ing to each ROM location may be output during two or more successive clock periods. This occurs when the number in the frequency setting register includes no 'ones' in the top N bits. On the other hand, at much higher frequencies, only a subset of ROM locations would be visited in one cycle of the output, a different subset usually applying in successive cycles. This gives rise to unwanted frequency components in the output; these may appear either as a few isolated spectral lines, or — for frequencies totally unrelated to the clock frequency — as a sea of low level spurs approximating to a raised noise floor. The cleanest output occurs when the selected frequency is a binary whole number, i.e. a power of 2 submultiple of the clock frequency; there are then no line spurs (other than harmonics of the output frequency), and the output is as pure as the clock frequency, possibly

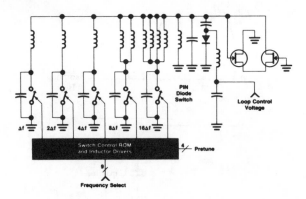

Figure 8.13 *This VCO used in the HP8662A synthesized signal generator is pretuned to approximately the required frequency by the microcontroller. The PLL error voltage therefore only has to tune over a small range, resulting in spectral purity only previously attainable with a cavity tuned generator, and an RF settling time of less than 500 μs. (Reproduced with permission of Hewlett-Packard Co.)*

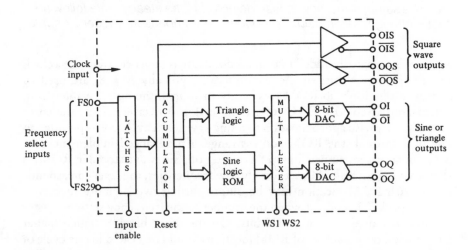

Figure 8.14 *SP2002 direct frequency synthesizer block diagram. (Reproduced by courtesy of GEC Plessey Semiconductors.)*

better, due to the division. At a small offset from such a frequency, close-to-carrier spurs will typically appear, the spacing being dependent upon the submultiple. For instance, at an output frequency offset by 1 kHz from $f_{clock}/4$, spurs would appear at ± 4 kHz.

The maximum output frequency from some DDS chips can be as high as one-third of the clock frequency or more, but in some designs (e.g. Figure 8.14) is limited by the architecture to $f_{clock}/4$. If working up towards the Nyquist frequency of $f_{clock}/2$, filtering will be required to suppress spurious outputs at image frequencies above the Nyquist rate. Figure 8.15a shows the output waveform of a DDS clocked at 400 MHz and set to provide an output frequency of 62.5 MHz, i.e. 5/32ths of the clock frequency. A different subset of levels (corresponding to ROM addresses) appears at subsequent cycles, the pattern recurring exactly after each fifth cycle. Thus, in the strict sense, the output is actually a 12.5 MHz signal, but with the fifth harmonic much stronger than the fundamental or any other harmonic, as can be seen on a spectrum analyser (Figure 8.15b). At more abstruse ratios than

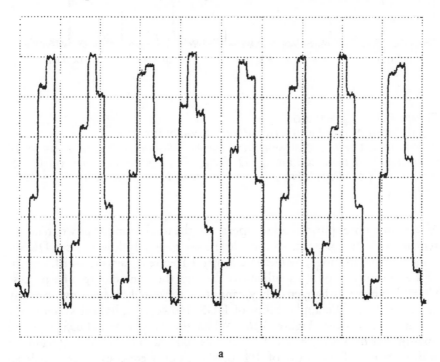

a

Figure 8.15 *Output of a direct digital synthesizer in the time and frequency domains.*
a Output of a DDS clocked at 400 MHz and set to $f_{out} = 62.5$ MHz. (The wiggles on the steps are an artefact of the digital storage oscilloscope used.)

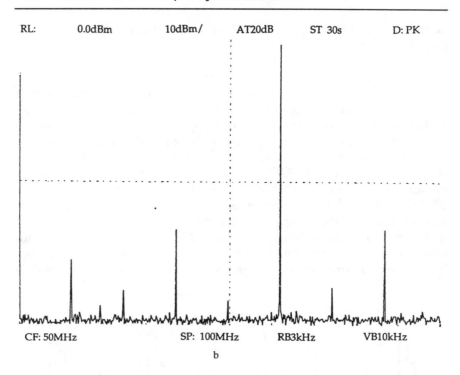

RL:　　0.0dBm　　　10dBm/　　AT20dB　　ST 30s　　D: PK

CF: 50MHz　　　　　SP: 100MHz　　RB3kHz　　VB10kHz

b

Figure 8.15 Continued
b Spectrum display (0–100 MHz) of waveform in a.
(Reproduced with permission from 'Direct digital synthesis, aspects of operation and application,' by D. May, IEE Electronics Division Colloquium on Direct Digital Frequency Synthesis, November 1991, Digest No. 1991/172.)

5/32, many more spurious lines appear, but the total spurious power tends to remain roughly constant, so their levels are generally lower. As a DDS is 'tuned' across its range, by incrementing the frequency setting word, various of the spurious outputs actually move through the wanted output frequency. Clearly, when this happens, they cannot be separated by filtering; in many cases this limits the applicability of DDS. However, a hybrid system may provide the answer (Figure 8.16). When the output of a DDS is set to one-quarter or less of the clock frequency, one can find frequency bands of width up to a few tenths of 1% of the clock frequency over which all spurious outputs are more than 80 dB down on the wanted output, although there may be spurs outside such a band. If the DDS operation is centred on 10.7 MHz, a highly selective crystal filter (such as used in PMR applications) can pick out a spurious free signal which may be set anywhere within the filter's bandwidth. With a reference frequency division ratio M of 5, the loop operates with a comparison frequency in excess of 2 MHz. This has

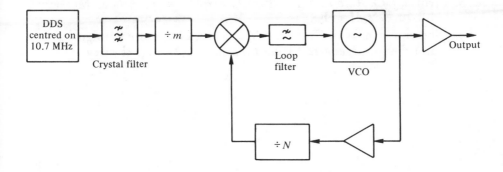

Figure 8.16 *Hybrid DDS/PLL synthesizer.*
(Reproduced with permission from 'Direct digital synthesis, aspects of operation and application,' by D. May, IEE Electronics Division Colloquium on Direct Digital Frequency Synthesis, November 1991, Digest No. 1991/172.)

two major benefits: firstly, a high loop gain may be retained up to a much higher frequency than normal, avoiding the rise in noise outside the loop bandwidth visible as 'ears' in Figure 8.10b and, secondly, the wide loop bandwidth results in very rapid settling following a change to a new frequency. The degree of resolution of the DDS, which typically has 30 or more bits in the frequency setting word, is so great that the synthesizer's output may be varied between the steps of the main loop in increments as small as 1 Hz or less. Note that this scheme provides its fine resolution by adjusting the frequency of the reference. The consequence of this is that the size of the fine loop steps is not constant, but proportional to the main loop divider ratio N. Thus, for a given synthesizer output frequency, the setting of the DDS must be calculated taking N into account, but this is no problem in a modern microprocessor-controlled design.

References

1. Robins, W. P. *Phase Noise in Signal Sources* IEE Telecommunications Series: 9 Peter Peregrinus
2. Scherer, D. Design principles and test methods for low phase noise RF and microwave sources. RF and Microwave Measurement Symposium, Hewlett-Packard
3. Burgoon, J. R. and Wilson, R. L. SC-cut quartz oscillator offers improved performance. *Hewlett-Packard Journal*, **32**(3), 20 (March 1981)
4. Gardner, F. M. *Phaselock Techniques*, John Wiley, New York (1966)

5. Truxal, J. G. *Automatic Feedback Control System Synthesis*, McGraw-Hill, New York (1955)
6. Sherer, Chan, Ives, Crilly and Mathiesen, Low-noise RF signal generator designs. *Hewlett-Packard Journal*, **32**(2), 12 (February 1981)

9
RF power amplifiers

This chapter covers the fundamentals of designing and testing RF power amplifiers. This differs from some other branches of RF design in that it deals with highly non-linear circuits. This non-linearity should be borne in mind when using analysis techniques designed for linear systems. The same problem also limits the accuracy of many computer modelling programs. This means that prototyping your designs is essential. With RF power electronics, thermal calculations become very important and this subject is also covered below — but before proceeding further, a word about safety.

Safety hazards to be considered

RF power amplifiers can present several safety hazards which should be borne in mind when designing, building and testing your circuits.

Beryllium oxide

This is a white ceramic material frequently used in the construction of power transistors, attenuators and high-power RF resistors. In the form of dust it is highly carcinogenic. Never try to break open a power transistor. Any component suspected of containing BeO that becomes damaged should be sealed in a plastic bag and disposed of in accordance with the procedures for dangerous waste. Do not put your burnt out power transistors in the bin, but store them for proper disposal.

High temperature

In a power amplifier, many components will get very hot. Care should be taken where you put your fingers if the amplifier has been operating for some time. When in the early stages of development, measurements on breadboarded PAs should be made as quickly as possible. The PA should be switched off between measurements.

Large RF voltages

High power usually means there are high voltages present, especially at high impedance points in the circuit. As well as the electric shock associated with lower frequencies, RF can cause severe burns. Take care.

First design decisions

The first design decision that should be made is that of operating class. For low power levels (less than about 100 mW) class C becomes difficult to implement and maintaining good linearity becomes difficult with class B. Unless the design requirement calls for a low-power transmitter that must be very economical with supply current then the best choice is usually class B for FM transmitters and class A for AM and SSB transmitters. At higher power levels (above 100 mW) the usual choice is class C for FM systems or other applications where linearity is not of concern, and class B for applications where good linearity is required, such as AM and SSB transmitters. The next choice is whether to design your own amplifier or buy a module. If considering an application in one of the standard communication bands using a standard supply voltage, then probably a module that will do the job can be found. Even if the use of a module is not contemplated, it is worth getting a price quote in order to obtain a benchmark to judge your proposed discrete design by. The choice whether to design your own or buy in an amplifier is dependent on the eventual production quantities of the project. If the quantities are small then the use of a module is probably the best choice as the small savings made in component cost per amplifier will be more than offset by the development costs of doing a discrete design. For large quantities then a discrete design should be costed and compared with the cost of a module. At the lower power levels it should be noted that most PA modules are of thick film hybrid construction resulting in a space saving that may be difficult to match with a discrete design. For high-power amplifiers that also require a high gain it is worth considering the use of a PA module as a driver for discrete output stage(s). The same module-versus-discrete decisions apply to the choice of harmonic filters. Harmonic filter modules are not as common as PA modules but there are plenty of small specialist filter design and manufacture companies that will design a filter to customer's specification. Because they specialize in filters they may be able to make the filters cheaper than your company can in-house.

Levellers, VSWR protection, RF routing switches

A VSWR protection circuit is required in many applications. This can be implemented using a directional coupler on the output of the PA. With a diode detector on the coupled port, the reverse power can be monitored as a dc level and used to initiate a turn-down circuit. The turn-down circuit works by reducing the supply voltage to the driver or output stage, or by reducing the drive power by some other means, for example by the use of a PIN attenuator. (The latter can also be used, under control of the output from the forward power monitor, for levelling, subject to overriding by the reverse power protection arrangements.) On MOSFET stages, another way of reducing the output power is to reduce the gate bias voltage. If the output stage is reasonably robust (i.e. the output device has power dissipation rating in hand) then the VSWR protection may just consist of a current limiter on the output stage. An approach that does not require such high dissipation rating devices in the control circuits is to use the current monitor to turn down the output power by one of the means outlined for the directional coupler approach, e.g. the current consumption of the output stage can be limited by reducing the supply voltage to the driver stage. The PA output may be routed via high-power PIN diode switches, to different harmonic filters, and/or to pads for providing reduced power operation.

Starting the design

Often the specification gives target figures for the output power and harmonic level from a combination of PA and harmonic filter. This leads to a chicken-and-egg situation in which the harmonic level from the PA needs to be known to specify the harmonic filter and the harmonic filter insertion loss is required to specify the PA output power. As a guide, start with the harmonic filter design for broadband applications, and start with the PA design in narrow band applications. For broadband matched push–pull stages, start with the assumption that the second harmonic is 20 dB below the fundamental and that the third is 6 dB below the fundamental. For broadband single-ended stages, use the starting assumption that the second harmonic is 6 dB below the wanted output. For narrow band designs a harmonic filter insertion loss of 0.5 dB is a reasonable starting point. These figures can be updated once some breadboarding has been done. The choice of a band-pass or a low-pass harmonic filter depends on several variables. If the operating frequency range is only a small percentage of the centre frequency then a band-pass design may well prove a better solution as a higher rejection can be achieved for a given order of filter. Band-pass filters usually involve a step up in impedance for the resonant elements and this can result in very high voltages being present. This aspect can limit the usefulness of band-pass designs at high power levels.

Low-pass filter design

(First a note about the definition of cut-off frequency. This is the frequency limit where the insertion loss exceeds the nominal pass-band ripple. With the exception of the Butterworth filter — a 0 dB pass-band ripple Chebyshev — and a 3 dB ripple Chebyshev, this is not the 3 dB point.)

Chebyshev filters

When the rate of cut off required is not too high and a good stop band is required, then a Chebyshev filter should be considered. The design method for these filters is based on look-up tables of standard filter designs. The values in these tables have been normalized for an input impedance of 1 Ω and a cut-off frequency of 1 Hz. Units are in farads and henrys. To choose which filter you require (for a given pass-band ripple), use can be made of the graphs giving attenuation at given points in the stop band, expressed as a multiple of the cut-off frequency. Once an order of filter and pass-band ripple has been chosen, the values can be taken from the tables and denormalized using the formulas in Figure 9.1.

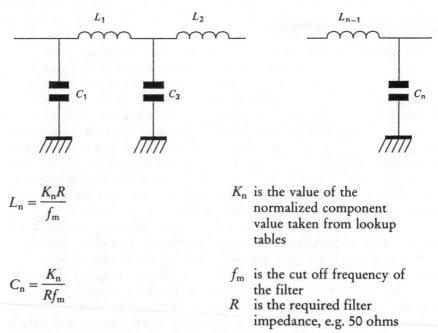

$$L_n = \frac{K_n R}{f_m}$$

K_n is the value of the normalized component value taken from lookup tables

$$C_n = \frac{K_n}{R f_m}$$

f_m is the cut off frequency of the filter

R is the required filter impedance, e.g. 50 ohms

Figure 9.1 *Filters: converting from normalized to actual values.*

Elliptic filters

The elliptic filter can achieve a sharper cut off than the Chebyshev but has a reduced stop-band performance. This filter type is best used where the PA has to work over a wide frequency range and therefore there is a requirement for a filter that cuts off sharply above the maximum operating frequency to give good rejection of the harmonics of the minimum operating frequency. The other application where an elliptic filter may be suitable is as a simple filter to reduce the second and third harmonics of a PA stage that already has a fair degree of harmonic filtering produced by a high Q output matching circuit. The design method is similar to that of the Chebyshev being based on standard curves and tables of normalized values.

Capacitor selection

There are three main dielectric types commonly used in capacitors for harmonic filters. They are mica, ceramic (NPO) and porcelain. Silvered mica capacitors can be used for harmonic filters in the HF spectrum. They tend to be larger than the ceramic and porcelain types and are not so common in surface mount styles. Their advantages are their availability in the larger capacitance values required for HF filters, and tight tolerance, tolerances as tight as 1% being readily available. NPO is a very common type and is readily available in surface mount. They are the cheapest of the three types. Their limitations are lower Q and lower voltage rating which limit their useful power range. Porcelain capacitors have a very high Q factor. Their RF performance is often better than documented by their manufacturers. These capacitors are usually used in the surface mount form to avoid lead inductance. The package sizes are not the industry standard 0805 or 1206 but come as cubes of side length 0.05 or 0.1 inches (1 inch = 2.54 cm). The 0.05 inch variety is usually rated at 100 V where as the larger size is rated at 500 V. These are the most expensive type of capacitor, costing about 20 times the NPO types. Larger (and even more expensive) types are available for very high power work with ratings of up to 10 A RF. When selecting a capacitor, points to consider are voltage rating, tolerance, availability in a reasonable size, and likely dissipation. The dissipation rating of a capacitor is often not given by the manufacturer so use the rating of a resistor of the same size as a guide. The dissipation in a capacitor can be calculated as follows. For shunt capacitors use the quoted Q figure to work out an equivalent parallel resistance and then calculate the RF dissipation in that resistance. For series capacitors calculate the RF current and calculate the dissipation in the equivalent series resistance (ESR).

Inductor selection

Depending on frequency, there are four main options for harmonic filters. Ferrite-cored inductors may be used at HF. The designer must be very careful that the ferrites are not saturated causing power loss and heating of the cores. Air-spaced inductors are to be preferred if at all possible. Air-spaced solenoid wound inductors can be used from HF to UHF and do not suffer from saturation effects. Losses are from radiation and resistance heating. Resistance heating includes losses due to eddy currents in any screening can that is used. Surface-mount inductors such as those made by Coilcraft can be used at VHF and UHF up to about 1 W RF output. These inductors suffer from poor Q, typically about 50, and wide tolerances (10%). For these reasons they should only be used where space is of prime importance. The vertically-mounted type on nylon formers provide a better Q (about 150 with screening cans) and a better tolerance of about 5%, trimmable if an adjuster core is fitted. They are available with or without screening cans. There is no rated dissipation given by the manufacturer's data sheet but practical harmonic filters have been found to get too hot to touch with an RF output power of 10 W, suggesting this to be the practical limit. If you wind your own coils then the best approach is to apply power and see how hot things get. If the enamel on the wire boils and spits, it is too hot. Printed spirals have the advantage of controllable tolerance and low cost. The disadvantage is they take up a large area of PCB and only have a Q in the range 50 to 100. An area with a height roughly equal to the radius of the spirals should be left clear above and below to avoid affecting the Q. The usefulness of printed spirals is limited to the VHF range. The final type is not strictly a true inductor, but a transmission line used as an inductor. This method is useful at UHF and higher. Conversion from inductance to line length is given by Equations 1 and 2 or can be read off a Smith chart. Z_0, the characteristic impedance, should be as high as practicable considering line loss and the effect of manufacturing tolerances. Wide low-impedance tracks can be made to a tighter tolerance than narrow high-impedance tracks.

Equation 1 Equivalent inductance of a transmission line shorted at one end

$$L = \frac{Zo \tan \Omega}{2\pi f}$$

Zo is the characteristic impedance of the transmission line

θ is the electrical length of the line in radians

Equation 2 Equivalent inductance of a short length of high impedance transmission line of impedance Zo in series with a load Z

$$L = \frac{(Zo^2 - Z_1^2) \tan \Omega}{2\pi f Zo}$$

Z_1 is the modulus of the load impedance

Discrete PA stages

With a bought-in module, much of the design process will have been done for you (though you may well still need to add harmonic filters). Therefore, most of the rest of this chapter is concerned with the design of discrete PA stages. One of the first decisions when designing an RF power amplifier stage is the choice of single-ended or push–pull architecture. A push–pull design will have the advantages of a lower level of second harmonic output and a higher output power capability. The lower second harmonic level makes broadband amplifiers simpler as each harmonic filter can be made to cover a wider pass band. The single-ended design has the advantage of fewer components, and is hence cheaper and requires less board space. Once the choice of architecture has been made, the next thing to consider is the load impedance presented to the transistor(s).

Output matching methods

There are two approaches that can be used to set the load impedance presented to the drain or collector of the RF transistor. Method A is to use the formula given by Equation 3 and collector capacitance data from the manufacturer's data sheet. The unknown quantity is V_{sat}; as a first approximation use 0.5 V for stages up to 5 W and 1 V above that. This is a very rough approximation, a more accurate figure is best obtained by experimentation. Method A ignores the presence of any internal impedance transformations that may be present. The practical implication is that inaccuracies increase as frequencies go up. Method B is to use large signal s-parameters or impedance data presented by the manufacturer of the transistor. (If no such data are available then method A should be used as a starting point.) It should be noted that these data are not the impedance 'seen' looking back into the device but the complex conjugate of the load impedance presented to the device which produces optimum performance for the output power and operating class stated. What this means is that the manufacturer has done some of your experimentation for you. If you want to use the device operating in a different way from that used by the

Equation 3

$$R_L = \frac{(V_{CE} - V_{sat})^2}{2P}$$

V_{sat} is the voltage drop from collector to emitter when the transistor is turned hard on

V_{CE} is the collector to emitter DC bias voltage

P is the output power

R_L is the output load resitance

manufacturer to characterize the device, you may have to resort to the equation given by method A. The manufacturer's output impedance data can be presented in several different forms. One method is to present tables or graphs (in Cartesian form) of the real and imaginary parts of the impedance. As an alternative, parallel resistance and capacitance tables or graphs may be given. It should be noted that the impedance data are in the form of a resistance in series with a reactance. Negative capacitance indicates an inductive impedance. The s-parameter data can be presented as tabulated values or a plot on a Smith chart. Once you have decided what impedance to match to, the next step is to decide how to implement the impedance conversion. Narrow band designs can be matched with lumped element or transmission line circuits as described in the input matching section below. For broadband designs, unless the collector load is close in value to the output impedance of the circuit (in which case a direct connection can be made with just a shunt inductor for DC supply and cancelling of collector capacitance), a broadband RF transformer will be required. The transformer places a limitation on the design by constraining the collector load to be an integer squared multiple or submultiple of the output impedance. This can be got around to a certain extent as discussed in the input matching section. If the impedance of any shunt reactive component is large compared with the resistive component, it can be ignored. If not, it can be tuned out as described in the input matching section. Broadband transformers are often based on a ferrite core. This should be large enough to avoid saturating the ferrite. The DC feed to the collector for single-ended stages should be taken via a separate choke to avoid adding to the magnetic flux in the transformer core. In push–pull stages the winding should be arranged such that the dc currents to each side cancel each others' flux contribution.

Maximum collector/drain voltage

The maximum voltage that will appear across the transistor is twice the maximum DC supply voltage. A transistor that has a breakdown voltage in excess of this figure should be chosen. RF power transistors have been optimized by the manufacturers to operate from one of the standard supply voltages. Choosing a transistor designed for a higher supply than is in use may give extra safety margin on the working voltage, but this will be at the expense of lower efficiency as the higher voltage device will probably have a higher V_{sat}. The standard supply voltages are 7 V, 12 V and 28 V. These standard supplies also tend to be used for power amplifier modules; in addition, 9 V is also used for some modules. The voltages relate to hand-held equipment, mobile equipment (vehicle mounted), and fixed (base station) equipment. The 28 V supply is also common in mobile (land and airborne) military equipment. Allowance must be made for supply voltage variations.

These can be severe, e.g. 18 to 32 V for a nominal 28 V dc supply, with even higher excursions if spikes and surges are teken into account. It may be necessary to stipulate a smaller range over which the power amplifier can be guaranteed to work to specification, with reduced output power capability at low voltage, and complete automatic shutdown in over-voltage conditions.

Maximum collector/drain current

Current consumption depends on the operating class. The easiest to calculate is class A as this is simply the bias current. For class B stages the peak current is given by Equation 4. For class C stages the peak current is a function of conduction angle. The smaller the conduction angle, the larger the peak current. The formula is given in Equation 5.

Equation 4

$$I_{peak} = \frac{2(V_{CE} - V_{sat})}{R_L}$$

Equation 5

$$I_{peak} = \frac{2\pi(V_{CE} - V_{sat})(1 - \cos \theta/2)}{R_L(\theta - \sin \theta)}$$ θ is the conduction angle in radians

Collector/drain efficiency

This is the efficiency of the output of the stage. It ignores power loss due to the input drive being dissipated and the power dissipated in biasing components. Collector/drain efficiency is the biggest factor contributing towards the overall efficiency of the amplifier stage. Class A is the least efficient mode, having a maximum theoretical efficiency of 50%. This figure ignores the effect of V_{sat}* which results in a practical figure less than the theoretical. As the conduction angle is reduced from the 2π radians of class A, the efficiency rises. The formula giving theoretical maximum efficiency is given in Equation 18. The derivation of this formula is given in Reference 1. A graph of this function is shown in Figure 9.2. From these you can see that the theoretical efficiency for a class B stage (conduction angle of π radians) is 78.5%. Class C is often quoted as a conduction angle of 120° ($2\pi/3$ radians) but in practice the conduction angle is difficult to control to any great accuracy. The theoretical maximum efficiency for a conduction angle of $2\pi/3$ is 89.7%.

*Collector saturation voltage, i.e. the lowest possible collector/emitter voltage for the given device and load.

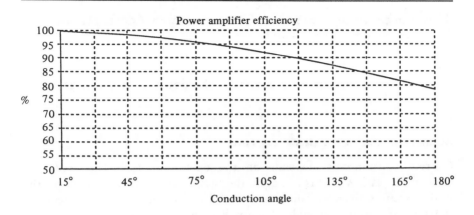

Figure 9.2 *Power amplifier efficiency.*

Power transistor packaging

There are many varieties of power transistor package and new ones are continually being developed. Figure 9.3 shows a selection of the most common types, categorized by dissipation rating. The two surface mount 1 W packages are relatively new. Use of the SO8 for RF power transistors is unique to Motorola but is a very common package for ICs. The SOT223 is made by Philips, Siemens and Zetex. This package looks like becoming an industry standard for 1 W devices in surface mount. Care should be taken when selecting a TO39 device as some transistors have the can connected to the collector, which can make construction more difficult as any heat sink used must be electrically isolated from the can. The ceramic studless package relies partly (as does the SO8) on the ground plane to conduct away heat from via the emitter leads: for this reason the emitter leads should connect directly to a large area of copper. In larger sizes one has the choice of flange-mounted or stud-mounted devices (stud-mounted devices also over-lap with the TO39 transistors). Devices of the highest dissipation rating are flange mounted. For flange-mounted devices there is the added choice of an isolated flange or one that is used as the ground connection. If you are using a PC board with a metal plate backing that doubles as heat sink and ground plane then the latter is the better choice. Otherwise the choice is dependent on mechanical arrangements. The isolated flange type is to be preferred in situations where the heat sink is not connected to the ground plane in close proximity to the RF power transistor. If designing a push–pull stage, then the dual transistor package is preferable as the stray inductance between the two devices is much less than that obtainable for two separate devices. It also

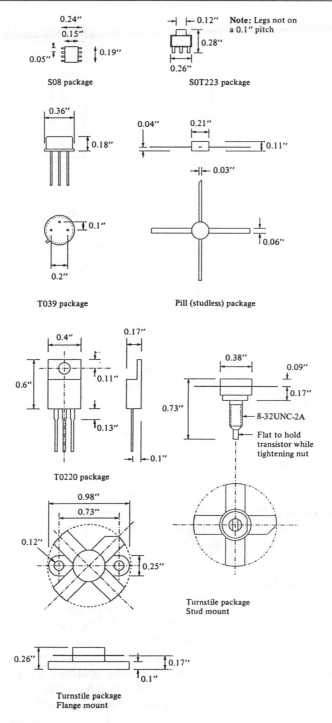

Figure 9.3 *Power amplifier packages.*

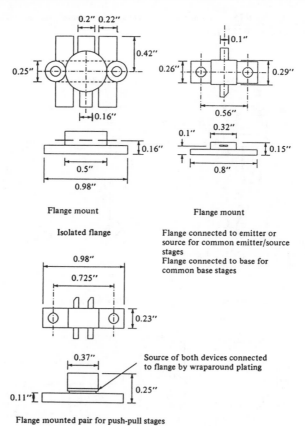

Figure 9.3 Continued

has the advantage that matched pairs are kept together. The devices designed for common base stages are usually only used for high power microwave amplifiers and are not discussed further here.

Gain expectations

The gain quoted by manufacturers in their data sheets is that measured in their test circuit. If operating the device in a different class, with a different load impedance, or with feedback or extra damping not included in the manufacturer's circuit then one can expect the gain to differ. If the device is characterized for class C operation but is being operated in class B then the gain will be higher (1 or 2 dBs). A move to class A operation will give even more gain. The choice of load impedance affects gain and efficiency. You may decide to sacrifice some gain in order to obtain higher efficiency or vice versa.

Thermal design and heat sinks

Thermal design is a very important part of RF PA design. The main source of heat will probably be the power transistor(s). To calculate the dissipation of a PA transistor the simplest approach is to calculate the difference between the power input and the power output. The power input is simply:

power input = DC collector/emitter voltage × DC collector current + input drive power

The power output is the RF power delivered into the output load. The maximum allowable transistor junction temperature and the thermal resistance from junction to case are usually given in the manufacturer's data sheet. Sometimes the manufacturer will quote a maximum dissipation and supply a derating curve instead. If this is the case the maximum junction temperature can be taken as the point on the derating graph where the allowable dissipation is zero. The thermal resistance can be taken from the slope of the graph. For those who are more accustomed to electrical design it helps to mentally transform the thermal circuit in to an equivalent electrical circuit. Power dissipated becomes current, temperature becomes voltage and thermal resistance becomes electrical resistance. As a minimum your thermal circuit will consist of a heat source (like current) and two resistors in series going to a constant temperature source. The first resistor is the device thermal resistance from junction to case, the second is the resistance of the heat sink to ambient, which is the constant temperature source. The resistances are usually in degrees Celsius per watt. The value for ambient should be the maximum expected and may need increasing to allow for solar heating if the equipment will be used outdoors. The circuit in a practical situation will probably be more complex with other heat sources summing in (e.g. more than one transistor bolted to the heat sink) and extra resistances for mounting brackets if they are used. Contact resistance can also play a significant part. To minimize this, mating surfaces should be as flat as possible and a very thin layer of heat sink compound used. With this information you will be able to calculate the maximum junction temperature achieved in the device for a particular heat sink. It is not a good idea to run the device continually at its maximum temperature as this will greatly reduce the reliability.

Biasing

MOSFETs are generally easier to bias in PAs then bipolar transistors as they are less susceptible to thermal runaway and do not draw current from their bias circuits. The disadvantage is that MOSFETs have a very wide tolerance on their gate threshold voltage. This means that either the circuit must be set up for each device fitted or some form of active bias control circuit be used. The simplest solution is a variable potentiometer, as shown in Figure 9.4. This can be adjusted to whatever bias current is required. The gate threshold voltage changes with temperature so this may be compensated for by adding a thermistor as shown. Figure 9.5 shows an example of an active bias circuit which needs no alignment to compensate for variation in the gate threshold voltage. This is a good solution for a class A stage which needs a constant current bias. Although the circuit is more complex, the extra components may well be paid for by reduced alignment costs. This circuit may also be used in a variable class mode if the set device current is less than that required for class A operation. In this situation the conduction angle becomes dependent on the drive power. For small drive powers the stage runs in class A. As drive is increased, the transistor starts to be turned off during part of the positive half of the output cycle. This distortion gives a dc component to the output waveform which tries to increase the current consumption. The control circuit will hold the current consumption at its set value by reducing the gate bias voltage. This will continue until the gate bias is at 0 V or the transistor starts to saturate on the negative half of the output cycle. A side effect of the changing conduction angle is that the gain is reduced with increasing drive. This will produce distortion of the RF envelope frequency components within the control loop bandwidth. As to whether this distortion is an advantage or disadvantage depends upon the application. Class A biasing for a bipolar transistor in the HF range can use a

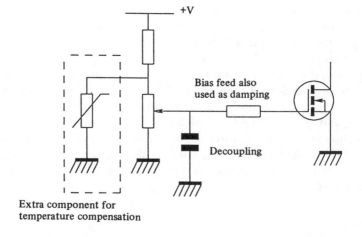

Figure 9.4 *Simple MOSFET bias circuit.*

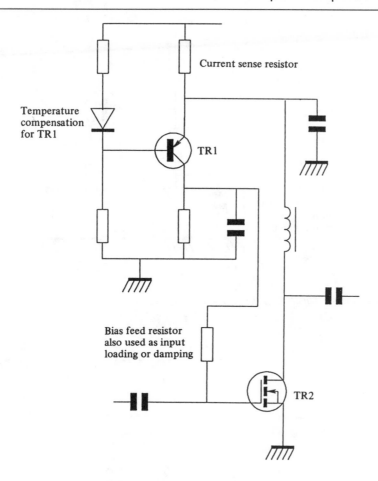

Figure 9.5 *Improved MOSFET bias circuit.*

bias circuit such as that shown in Figure 9.6. This can be temperature compensated as shown. The layout should be designed to minimize the length of the RF path from the emitter to ground. Any inductance in series with the emitter will reduce the gain of the stage and may compromise the stability. An alternative which can be used if a stabilized supply is in use is shown in Figure 9.7. This method has the advantage of having the emitter connected directly to ground, minimizing stray inductance and allowing use at higher frequencies. A variation of the active bias circuit used for MOSFETs can be used as shown in Figure 9.8. This is much less dependent on supply voltage. A simple Class B bias circuit is shown in Figure 9.9. Close thermal coupling between the diode and RF transistor is necessary to ensure thermal stability. When there is no RF drive the bias current in the

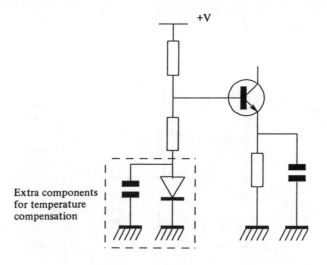

+V

Extra components
for temperature
compensation

Figure 9.6 *Simple bipolar bias circuit.*

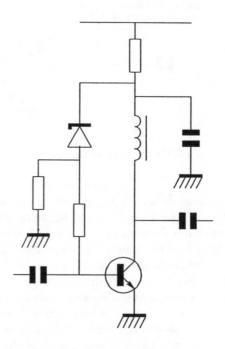

Figure 9.7 *Improved bipolar bias circuit (1).*

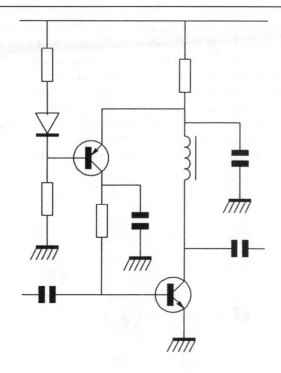

Figure 9.8 *Improved bipolar bias circuit (2).*

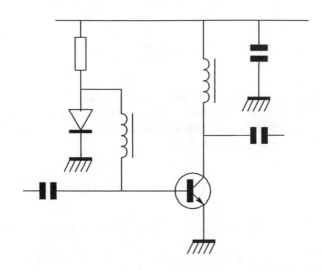

Figure 9.9 *Simple bipolar bias circuit for class B.*

transistor will be approximately the same as that flowing through the diode. When drive is applied, the base current will increase. This will cause less current to flow in the diode and hence the bias voltage to drop. It is up to the designer to ensure that the diode current does not drop to zero when the drive is at its maximum if he or she does not want the stage to go into class C operation, with the resulting loss of gain and envelope distortion. Closed loop bias control is not possible as the current is inherently drive dependent. The simplest form of class C bias is shown in Figure 9.10. A resistor can be put in series with the choke which will negative bias the base emitter

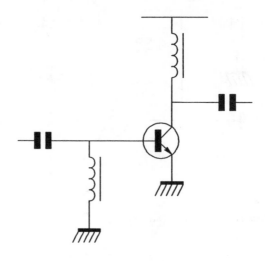

Figure 9.10 *Simple class C bias circuit.*

junction using the base current. If you do use this method, care is required to make sure that the reverse breakdown voltage of the base emitter junction is not exceeded even under worst case conditions. The maximum reverse base emitter voltage is given in Equation 6.

Equation 6

$$V_{peak} = \sqrt{2P_{in}R_{in}} + R_bI_b$$

P_{in} is the input power to the device

R_{in} is the input resistance of the transistor

R_b is the base bias resistor

I_b is the base bias current

Feedback component selection

Feedback on a PA stage usually consists of a resistive or complex impedance connected between the drain/collector of the transistor and the gate/base or, less commonly, a resistor between the emitter/source and ground. The latter is to be avoided above HF use and above medium power as the resistance required is usually very low and can easily be swamped by circuit strays, causing a roll off in high frequency gain and power output. Drain to gate feedback is often used to aid stability and control gain in MOSFET stages. Consider the circuit shown in Figure 9.11. The addition of the drain to gate feedback resistor has several effects:

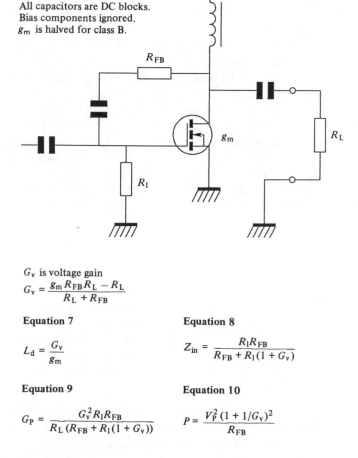

Note

All capacitors are DC blocks.
Bias components ignored.
g_m is halved for class B.

G_v is voltage gain

$$G_v = \frac{g_m R_{FB} R_L - R_L}{R_L + R_{FB}}$$

Equation 7

$$L_d = \frac{G_v}{g_m}$$

Equation 8

$$Z_{in} = \frac{R_1 R_{FB}}{R_{FB} + R_1(1 + G_v)}$$

Equation 9

$$G_P = \frac{G_v^2 R_1 R_{FB}}{R_L (R_{FB} + R_1(1 + G_v))}$$

Equation 10

$$P = \frac{V_P^2 (1 + 1/G_v)^2}{R_{FB}}$$

Figure 9.11   *Drain/gate feedback (resistive).*

(a) It reduces the drain load to that shown in Equation 7.
(b) It reduces the input impedance as in Equation 8.
(c) Because of (a) and (b), it reduces the gain to that shown in Equation 9.
(d) Due to the power dissipated in the feedback network, the efficiency is reduced. The power dissipated in the feedback resistor is given in Equation 10.

The gain figure from Equation 9 ignores the effect of any reactive components in the circuit, including those within the transistor. The device's drain to gate capacitance acts in parallel with the external feedback resistance and can be considered as part of a complex feedback network. Adjustments to the circuit can be made to compensate for the effects of the feedback capacitance over a limited frequency range. If the reactance of the feedback capacitance is large compared with the feedback resistor then an inductor in series with the resistor may be all that is required for compensation. A recommended inductor value is given by Equation 11. The resulting network is a two-pole low-pass terminated by the resistor. Depending on the Q of the network, the circuit may produce a gain peak at the value of F_{max}. When the reactance of the feedback capacitance approaches that of the feedback resistance, then the network in Figure 9.12 can be used. The value of the inductor is two times that given in Equation 11. The capacitor value is the same as that of the feedback capacitance of the transistor. The choice of feedback network is dependent on what degree of gain flatness is required. For push–pull stages there is another way of reducing the effect of feedback capacitance. This is shown in Figure 9.13. This method should be used with care as it effectively introduces positive feedback. The value of the feedback capacitance can vary greatly between samples of a particular device type.

Equation 11

$$L = \frac{CR_{FB}^2}{1 + (R_{FB}2\pi F_{max}C)^2}$$

C is the feedback capacitance of the transistor
R_{FB} is the feedback resistor
F_{max} is the maximum operating frequency

Unfortunately transistor manufacturers rarely quote minimum feedback capacitance, only typical and/or maximum. For many devices the maximum figure is twice the typical. This suggests, assuming an even distribution, that a good minimum figure is half the quoted typical or a quarter the maximum.

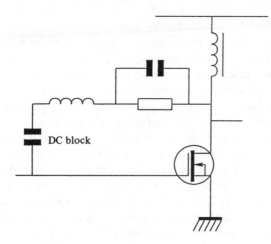

Figure 9.12 *Complex feedback.*

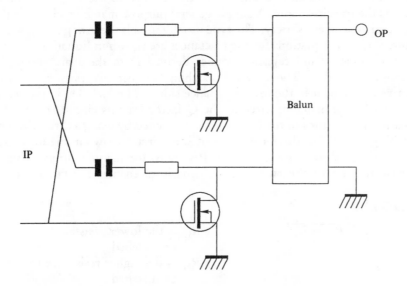

Figure 9.13 *Cross neutralization.*

In order not to compromise the stability of the circuit, the cross-connected capacitors should not be larger than this minimum figure. The value of the resistors to be used is best found out by experimentation. They are there to maintain high frequency stability.

Input matching

When discussing a general class of devices, such as bipolar transistors, the discussion has by necessity to be very vague. There is also a large number of solutions to any particular matching problem. Despite all this, some general comments follow, concerning the type of matching circuits required in PA input matching, and how to design them. In general the input impedance of a bipolar PA transistor is in the order of a few ohms resistive plus a reactive component. At lower frequencies the reactive component is capacitive, and at higher frequencies it is inductive. The cross-over point is in the mid VHF band. The resistive component becomes lower as the power of the stage goes up. At VHF and above, particularly in the higher power devices, impedance matching circuits are included inside the transistor package. These do not usually match direct to 50 Ω, but raise the very low input impedance of the transistor to an impedance which, though still lower than 50 Ω, is much easier to match. The typical construction of such matching is shown in Figure 9.14. The internal matching shunt capacitor has the advantage over external circuits in that one end is directly attached to the same grounding point as the transistor chip. A simple general purpose matching circuit is the two-lumped element variety. The type usually used is the low-pass shown in Figure 9.15. The equations for the reactances are shown in Equations 12 and 13. The inductor and capacitor values derived from them are shown in Equations 14 and 15. These are for matching between two resistances. Any reactive component in the low impedance side can be included in the series reactance of the matching circuit. The Q factor for this circuit is given by Equation 16. Control of the Q factor can be gained by using a three-element matching circuit. The three-element matching circuit shown in Figure 9.16 is commonly used as a test circuit by PA transistor manufacturers. This is because the use of the two variable capacitors enables the circuit to be

Equation 12

$$X_{\text{Series}} = \sqrt{R_L R_H - R_L{}^2}$$

R_L is the lower resistance to be matched

R_H is the higher resistance to be matched

Equation 13

$$X_{\text{Shunt}} = R_H \sqrt{\frac{R_L}{R_H - R_L}}$$

Equation 14

$$L = \frac{\sqrt{R_L R_H - R_L{}^2}}{2\pi f}$$

Equation 15

$$C = \frac{1}{2\pi f R_H} \sqrt{\frac{R_H - R_L}{R_L}}$$

Equation 16

$$Q = \sqrt{\frac{R_H - 1}{R_L}}$$

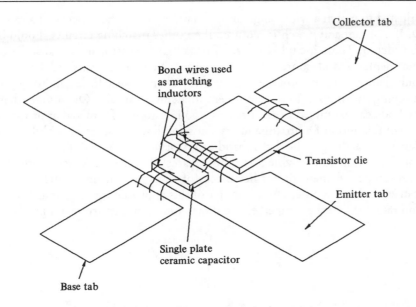

Figure 9.14 *Transistor with internal input matching.*

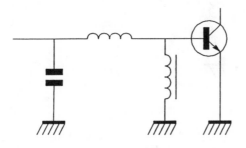

Figure 9.15 *Two element matching circuit.*

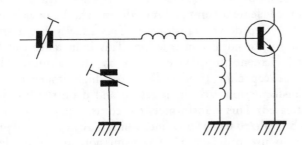

Figure 9.16 *Three element matching circuit.*

adjusted to match a wide range of impedances, but at the expense of a raised Q. If a broadband match is required then other matching circuits should be considered. These include the use of broadband transformers, transmission line elements and more complex lumped element circuits, such as the four-element circuit shown in Figure 9.17. There is very little gain to be had in going beyond a four-component matching circuit. Of course these methods can be mixed as required. A good example of a mixed approach is the combination of a broadband transmission line transformer with lumped element matching. The broadband transformer is limited to impedance transformation ratios which are the squares of integers. When combined with lumped element or further pieces of transmission line matching, this restriction is overcome. The advantage of this approach for large transformation ratios is that the lumped element matching can start from an impedance

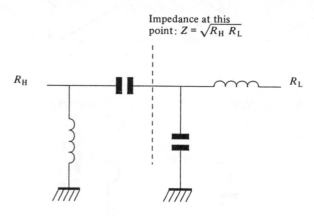

Figure 9.17 *Four element matching circuit.*

much closer to that desired and therefore have a much lower Q. Often the lumped element matching components can be included within the broadband transformer. Practical RF transformers are not ideal and therefore have strays that can be modelled as lumped elements. These strays can be used as part of the lumped element component of the match. As an example of this, consider the 4 : 1 step-down transformer. This usually has a small series inductance due to non-ideal construction. This inductance can be turned into a lumped element impedance match by the addition of a shunt capacitor. If the capacitor is placed on the high impedance side, the impedance transformation ratio is increased and if on the low impedance side, it is decreased. This transformer if used as a step down from 50 Ω would ideally be realized using 25 Ω line, which may not be very practical. A useful trick is to use ordinary 50 Ω transmission line, thus deliberately increasing the series stray inductance of the transformer, hence increasing the range over which the transformation ratio can be adjusted. The amount of extra inductance created by this trick is obtained using Equation 2. In practice the other contributions such as connecting leads add significantly to this figure so the final arrangement should be built, measured and adjusted before use. There are many other areas where a practical design will probably be forced to depart from ideal RF construction. The trick of good RF design is to use the strays caused by construction limitations to one's advantage. The limiting factor for lossless broadband matching is the Q of the input impedance of the device. To go beyond this limitation some gain must be sacrificed by the inclusion of resistors external to the device to reduce the Q, or the acceptance of some mismatch. Broadband MOSFET input matching is an extreme example of using resistors to limit the Q of the input match. In this case a shunt resistor is used to provide the majority of the input load. A MOSFET transistor's input impedance is mainly capacitive and therefore cannot be broadband matched without this shunt resistor. Feedback resistors may also play a significant part in defining the input impedance, and in some circuits form the main part of the input impedance.

Stability considerations

Stability is a very important subject in power amplifier design. It can also be very hard to get right. MOSFETs usually display better stability than bipolar transistors. Due to the non-linear processes present, the stability criteria based on s-parameters (Appendix 2) do not always predict potential oscillations. A bipolar transistor has a reverse biased diode as the collector base junction. This behaves as a varactor diode causing frequency multiplication and division. Frequency division is a common problem in broadband class C stages, and is a symptom of being overdriven or having not enough output voltage available. A MOSFET has a parasitic diode between drain and substrate which can show similar effects. The frequency division aspects are particularly bothersome, as the gain of the devices is usually higher at the lower frequencies. The best way to assess stability is by extensive testing. Stability problems are best overcome by careful layout and the addition of resistive dampers. A base/gate damping resistor should be included from the outset. This is required to limit the Q of any resonance with bias chokes and matching transformers. As an alternative, the damping resistor can be used as a bias injection route, saving on one inductor; however, this is not recommended for bipolar class C stages as the base current drawn will probably cause too much reverse bias of the base emitter junction. As a general rule of thumb, use a resistor value that is four times the base/gate input impedance. If you can get away with damping just at the input, then no output damping should be used as this tends to waste output power. If the oscillations occur at a frequency lower than the required operating range then frequency selective damping on the input and/or output as shown in Figure 9.18 may be used without dissipating too much of the wanted output

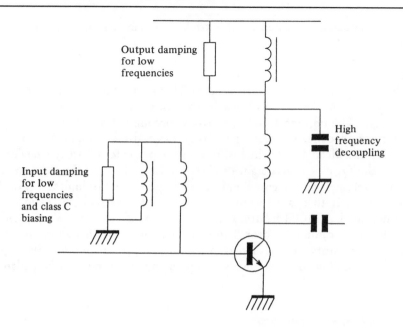

Figure 9.18   *Damping circuits to improve usability.*

power in the damping resistor. A technique widely used to stabilize MOSFET stages which have a very large LF gain is to use feedback resistors. Even if they are too high to affect the gain at the operating frequency, they may well successfully prevent oscillations at lower frequencies.

Layout considerations

As a general rule, the higher the frequency and the higher the power, the less you can get away with. Layout should have regard to the impedance at each part of the circuit in question. For low impedance parts of the circuit, minimizing stray series inductance should be of prime concern. For high impedance parts of the circuit, minimizing stray shunt capacitance should be the prime concern. Earth returns, particularly those carrying high RF currents, should be made as short as possible. Sources of stray inductances include component leads, connecting wires to coaxial lines, and lengths of tracking with a characteristic impedance higher than the operating impedance at that point. Sources of stray capacitance include tracking spurs on the PCB and lines of characteristic impedance lower than the operating impedance of the circuit at that point.

Construction tips

The combined requirements of good heat sinking and good RF layout practice often lead to the requirement for a large metal plate associated with the PCB. If it is necessary that the heat sink also provide a good RF earth, the logical extension of this is a thick metal plate bonded to the PCB. The metal plate forms both part of the heat sink and the ground plane. When the heat sink and PCB are separate, repeated assembly and disassembly should be avoided as this can mechanically overstress the bolt-down components. Stud-mounted transistors should not be soldered to the PCB until they have been bolted down to avoid stressing the leads.

Performance measurements

Power output is usually measured with a power meter. Power meters can be split into two broad groups: those based on thermal heating in a load and those based on diode detectors. Both types will give false readings in the presence of high harmonic levels. The thermal type indicates the total power, including harmonics. The error E due to a second carrier such as a harmonic is shown in Equation 17. If only one harmonic is at a significant level and that level relative to the fundamental is known, then this formula can be used for calculating a correction factor. The diode detector types can indicate high

Equation 17

$$E = 10 \log(1 + 10^{-d/10})$$

d is the difference between the signal to be measured and the 2nd signal, measured in dBs.

Equation 18

$$\eta = \frac{\theta - \sin \theta}{2(2 \sin(\theta/2) - \theta \cos(\theta/2))}$$

θ is the conduction angle in radians

or low depending on the phase of the harmonics relative to the fundamental. Spectrum analysers can be used to measure power without readings being affected by harmonic levels; however, absolute power measurements with spectrum analysers are not as accurate as those by thermal power meters such as the HP436A. The harmonic output of a PA stage is simply measured using a spectrum analyser, with a suitable high-power attenuator to bring the carrier power down to a safe level for the spectrum analyser. When the item under test is a PA and harmonic filter combination, the harmonic output may be lower than that produced internally in the spectrum analyser being used to make the measurement. To avoid this problem a test set-up as shown in Figure 9.19 can be used. This uses the notch filter to remove the fundamental of the transmit spectrum, leaving the harmonics to be measured

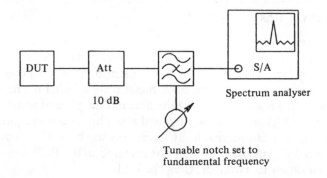

Figure 9.19 *Testing a PA/harmonic filter combination.*

with the spectrum analyser. The attenuator is required to present a reasonable load to the circuit under test. For the higher order harmonics a practical notch filter may be excessively lossy. If this is the case then a high-pass filter can be used in place of the notch for these measurements. Stability into mismatched loads is an important consideration. In the real world, exactly matched loads do not exist — a practical PA will have to tolerate some mismatch. The stability of a PA design will need testing into the worst case VSWR at all phase angles. In non-linear circuits, supply voltage, temperature, and drive power also will have an effect on stability. Testing the many permutations of these variables is a long and time-consuming job, but for a good PA design it cannot be avoided. A method of presenting a variable phase mismatch and monitoring the output spectrum is shown in Figure 9.20. The phase shifter should be able to present a load that traverses the entire outer ring of the Smith chart at the operating frequency (from short circuit to open circuit and back again). This can be done with a 'trombone' (a variable length coax line or 'line stretcher') terminated with a short circuit, or a lumped element line stretcher as described in Reference 2. Unlike linear circuits, the input impedance of a PA stage is a function of drive level and supply voltage. Consequently, measurements of input impedance must be made at the design drive level applying in actual use. When the device under test is an unmatched transistor or the existing matching circuit does not give a good match, then the drive from the measurement system may need to be higher than the nominal drive requirement of the circuit in order to get good results. The drive requirements are often beyond the output power capabilities of a network analyser. A typical test set-up for measuring input impedance is shown in Figure 9.21. The device under test should always be tested in to its working load, with any output matching circuits in place. With many devices the mismatch between unmatched input and test system is so great that it is not practical to make up for drive loss by just increasing the drive from the test system. In

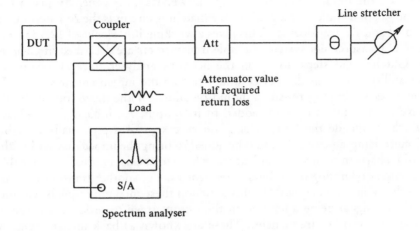

Figure 9.20 *Testing a PA into high load VSWRs.*

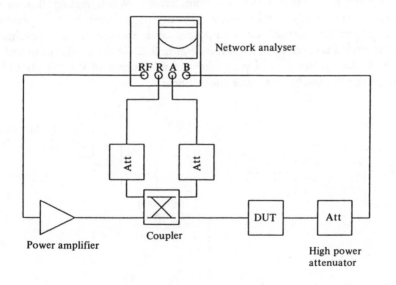

Figure 9.21 *High level testing of input VSWR.*

these cases some form of input matching will be needed from the outset. If these matching circuits are characterized on their own beforehand then readings can be translated to get the actual input impedance of the device. Because the input impedance of high-power stages is generally just a few ohms, a good choice for a preliminary matching circuit is the 2 : 1 step-down broadband RF transformer. This gives a working impedance of 12.5 Ω from a 50 Ω measurement system. Suitable transformers are described in Chapter 3. Glitches and steps down on the network analyser trace are a sign of instability, either in the device under test or the measurement system. In these cases damping resistors should be added or the drive source should have a low value attenuator added to its output. An indicated impedance which is outside the Smith chart is a sure sign of a potentially-unstable circuit; damping circuits should be added to bring the impedance within the Smith chart. In service an amplifier may have to coexist in proximity to other amplifiers operating on different frequencies, e.g. another transmitter sharing the same antenna mast. In this situation these incoming signals will mix with the signal being amplified in the output stage to produce a range of products on other frequencies. These are known as back intermodulation products or reverse intermods. The level of these intermodulation products will have to be measured to check that they are not going to be large enough to interfere with other radio communications. When testing this in the laboratory one needs to take precautions against intermodulation products being generated in the test equipment and corrupting the results. A recommended test set-up is shown in Figure 9.22. If the levels produced are too high then either a band-pass filter on the output of the PA should be used or the PA should be made more linear.

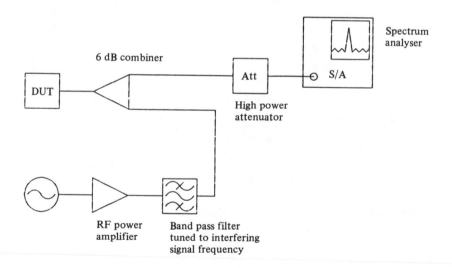

Figure 9.22 *Reverse intermodulation testing.*

References

1. Smith, J. *Modern Communication Circuits*, McGraw-Hill, New York
2. Franke, E. A. and Noorani, A. E. Lumped-constant line stretcher for testing power amplifier stability. *RF Design*, March/April, 48–57 (1983)

10
Transmitters and receivers

The previous chapters have covered all the circuit functions used in transmitters and receivers, but when putting them together into a TX or RX equipment, or indeed a T/R (transmitter/receiver, e.g. Figures 10.8 and 10.9), then certain additional considerations arise. These are considered below.

Figure 10.1a shows a block diagram of a transmitter, such as might be used for VHF FM broadcasting. The baseband signal consists of the programme input material, speech or music, nowadays often in stereo. Baseband signal processing produces the mono-compatible sum signal, the stereo difference signal which is modulated onto a suppressed subcarrier, and the stereo pilot signal at half the frequency of the subcarrier. Often also, RD (radio data) information at a low bit rate is modulated onto an additional subcarrier. This carries a variety of information such as station identity, other frequencies on which the same programme can be received (useful for auto-searching FM receivers in cars), etc. The composite baseband signal is modulated onto a carrier at a suitable IF frequency such as 10.7 MHz and then, after filtering to the final bandwidth, translated in a mixer stage to the final transmit frequency. In the USA, the serasoidal modulator was at one time popular, but this has a maximum phase deviation less than ±180°. Frequency multiplication was therefore necessary to obtain the required deviation, making it difficult to achieve an acceptable signal to noise ratio even with a mono signal. In a broadcast transmitter, the transmit frequency is seldom if ever changed, so tuning arrangements are much simpler than those commonly found in receivers. However, sophisticated protection arrangements for safety purposes are necessary, including interlocks to prevent the equipment being accidentally powered up whilst personnel are servicing it, and trips to protect the PA in the event of an antenna fault, etc. In one sense, a good transmitter is easier to design than a good receiver, since the only signal it has to handle is the wanted signal. This is especially true of a transmitter working over only a fairly narrow percentage bandwidth such as the 88–108 MHz VHF FM broadcast band, as it is then easy to arrange that

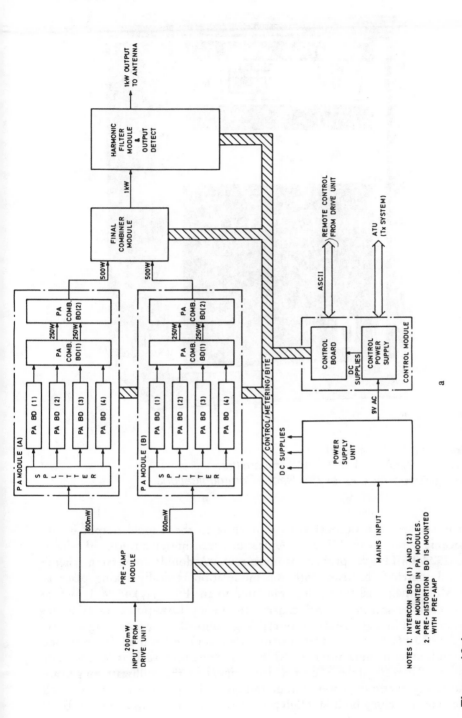

Figure 10.1
a Block diagram of a modern 1 kW HE transmitter.

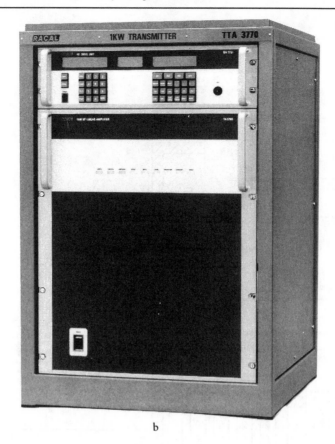

b

Figure 10.1 Continued
b The Racal TTA3770 1 kW transmitter.
(Reproduced by courtesy of Racal Communications.)

no mixer spurious outputs fall on or close to the wanted output in the transmit band. In an HF communications transmitter covering the band 1.6–29.999 MHz, the problem is more acute. A double conversion scheme would therefore be used with the modulation typically taking place at 1.4 MHz, the signal then being translated to an IF of (say) 45 MHz before down conversion to the final transmit frequency. Low-power UHF transmitters used in walkie-talkies, portable telephones, etc., operating in parts of the 470–960 MHz spectrum usually use complete PA modules from one of the leading manufacturers of RF power transistors, such as Motorola or Philips. These modules accept a drive signal in the milliwatt range, are available in various power output ratings and are ready set up with all interstage matching built in. High power transmitters in this band, e.g. Band

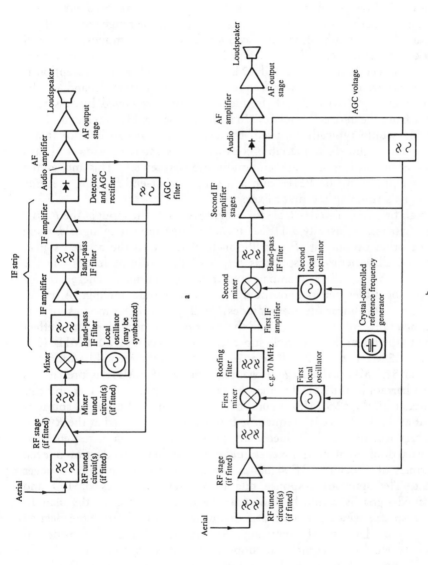

Figure 10.2 *a Single-conversion superhet. Several filters may be used throughout the IF strip. b Double-conversion superhet, with synthesized first local oscillator and second local oscillator both crystal reference controlled.*

IV/V TV transmitters, use valve PAs, although solid state transmitters are currently pushing up to a power level of kilowatts.

Figure 10.2a and b shows single and double superheterodyne receiver block diagrams, such as might be used in a quality short-, medium- and longwave AM radio and an HF communications receiver respectively. In the AM single superhet, the IF frequency is typically in the range 455–470 kHz with an IF bandwidth of as little as 5 kHz, allowing a modest degree of rejection of stations on adjacent channels (medium wave channel spacing is at 9 kHz intervals in Europe and 10 kHz in USA). However, reception is usually restricted to the lower frequencies in the short waveband, as the image frequency (twice the IF frequency) is only removed by less than 1 MHz from the desired frequency. In a single superhet HF receiver an IF of 1.4 MHz would typically be used, but even this leaves an inferior image performance. Therefore a double conversion system is nowadays always employed in professional HF communications receivers. This moves the image frequency to the VHF band and simple front-end filtering prevents such signals reaching the first mixer.

A high first IF is also desirable for other reasons. If the input at the R port of the first mixer (usually a DBM) includes large unwanted signals, there may be other outputs at IF in addition to that due to the wanted signal. These are all varieties of 'spurious response' due to imperfections in the DBM which the mixer manufacturer tries to minimize. There are for example possible spurious outputs due to harmonic mixing. A mixer containing non-linear devices (diodes), will produce harmonics of the frequencies present at its inputs, and these harmonics themselves are in effect inputs to the mixer. So if a single superhet HF receiver with a 1.4 MHz IF is tuned to 25 MHz, the LO will be at 26.4 MHz and the second harmonic of this is at 52.8 MHz. If a large unwanted input at 25.7 MHz is present, its second harmonic at 51.4 MHz may be produced within the mixer and this will beat with the 52.9 MHz second harmonic of the LO to give a spurious output at the 1.4 MHz IF frequency. If the mixer is balanced at the R port, the effect will be greatly reduced but, in practice, not eliminated entirely. The usual double balanced mixer should not result in the production of even harmonics of either the RF signal or the LO, but mixer balance is never perfect. The spurious response due to second harmonics of LO and unwanted signal is variously known as the '2:2 response' or the 'half IF away response' since it occurs at a frequency removed from the desired frequency by half the IF frequency. An impractical degree of front-end selectivity would be required to suppress this response to a level where a 100 mV unwanted signal would not drown a 1 μV wanted signal. Further, a double balance mixer offers no such enhanced rejection to the 3:3 response, removed from the tuned frequency by only one-third of the IF frequency, or other odd order responses. This type of receiver spurious response falls off rapidly as higher and higher order harmonics are involved. It can thus be avoided virtually completely by using a double superhet configuration with a first IF well above 30 MHz, since the harmonic orders involved would then be very high. Possible responses at the IF, image and at frequencies as described above are all examples of external spurious responses or 'spurs'. Most receivers, even professional communications receivers, will have one or more internal spurs. These are frequencies at which there is an apparent CW output even with the antenna input terminated in a resistive load. They are due to spurious spectral lines occurring in the synthesizer and/or interactions between the first and second local oscillator and the frequency standard. Other possibilities are harmonics of the clock frequency of the microcontroller included in all modern receivers.

A superhet is troubled by other types of spurious responses, of which intermodulation is one. Imagine the receiver is tuned to a weak wanted signal and that there are two large unwanted signals, removed by +100 kHz and +200 kHz from it. The lower of the two third-order intermodulation products of the unwanted signals will fall on the wanted frequency: the formation of intermodulation products due to circuit non-linearity is covered in Chapter 5. In a professional HF communications receiver, e.g. Figure 10.7, the third order intermodulation performance is usually specified with unwanted signals offset from the tuned frequency by ±20 and 40 kHz, at which spacing there will be no assistance from any front-end tuning. However, second order intermodulation products will not be a problem except in a 'wide open' receiver with no front-end tuning of any description: a high quality HF receiver will usually have either a tuned front end or a bank of nine sub-octave band-pass filters covering the 1.6–30 MHz band. The appearance of high dynamic range double-balanced mixers led in the 1970s to a rash of wide open HF receivers, but with the ever heavier use of the HF band and the resulting mayhem against which receivers have to work, the true worth of a tuned front-end is again recognized.

Two other headaches for the receiver designer are cross-modulation and blocking (desensitization). In the former, the envelope modulation on a large unwanted off-tune signal becomes impressed on a smaller wanted signal and cannot therefore be removed by any subsequent filtering. Blocking consists of a reduction of gain to the wanted signal, caused by a large unwanted off-tune signal. Cross-modulation and blocking are usually specified for an unwanted signal offset of 20 or 30 kHz. Like intermodulation, they would not occur in a receiver in which all stages up to and including the final bandwidth defining second IF filter were perfectly linear. It is for this reason that most of the gain is provided in the second IF stages following the final bandwidth filter — by that time the only signal present is, it is to be hoped, the wanted one. Keeping the gain as low as possible in the earlier stages minimizes the size of any large unwanted signals in those stages, minimizing the effect of their inevitable slight non-linearity. However, sufficient gain must be provided to compensate for attenuation in tuned circuits, mixers, etc., so that the signal to noise ratio of a small wanted signal at the input to the receiver does not become noticeably worse at the receiver's output. As the level of the wanted signal increases, the receiver's gain must be turned down so as not to overload the last IF stage and/or detector. The operator can do this using the manual RF gain control if provided, but usually it is the job of the AGC (automatic gain control) circuitry, which is 'scheduled' so as to maintain the best signal to noise ratio for the wanted signal. The gain at the back end of the second IF amplifier strip is turned down first, to approximately unity. Then earlier stages are successively turned down, until eventually the gain of the RF stage (if fitted) is turned down, or alternatively a voltage controlled attenuator preceding it is brought into operation. AGC which is scheduled in this way provides better performance than winding down the gain of all controlled stages in parallel, or applying full AGC to the IFs and half AGC to the RF stage. It is arranged that the final IF stage is capable of driving the signal and AGC detectors to full output even at maximum gain reduction, either by limiting the gain reduction of that stage or by not controlling it at all. Compared to manual RF gain control, AGC has of course the advantage that it will continually adjust the receiver's gain to compensate for variations of the strength of the wanted signal due to fading. Typically, sufficient gain is provided in the AGC loop to keep the variation in output signal level to 5 dB or less for a change in input level of 100 dB. AGC is not without its problems: AM signals such as broadcast stations on short wave (and on medium wave, after dark) may suffer selective fading of the carrier, leaving the sidebands unaffected. The AGC will increase the receiver's gain leading to a large increase in the audio output level, which will moreover be grossly distorted, since in the absence of the carrier, the modulation index is way in excess of 100%. The attack, hold and decay times of the AGC loop will be set to appropriate values for the mode of reception selected. Thus short time constants will be used for AM reception, where there is (normally!) a carrier providing a continuous indication of received signal strength, but much longer hold and decay times are used in SSB mode. Here, the absence of any carrier results in the disappearance of the signal during pauses in speech: a rate of gain recovery (decay) of 20 dB/s is typical. AGC action generally starts at or a few decibels above the receiver's rated sensitivity level, which for an HF receiver in SSB mode would typically be 1 μV EMF for a 10 dB SINAD (signal to noise-plus-distortion) ratio. This corresponds to an NF (noise figure) of about 15 dB, which is usually perfectly adequate for the HF band, where atmospheric and man-made noise levels are very high most of the time. Some HF receivers boast an NF of 10 dB or even lower: there are rare occasions where this can be useful such as when constrained to operate with a grossly inefficient aerial. An example is operating from a nuclear bunker where the antenna is a very short blast-proof whip or is even buried. Some HF receivers have a stage of RF gain which can be bypassed, or switched in to obtain a lower noise figure when no large signals are present, e.g. on a merchant ship alone in the midst of the ocean.

The other main class of receiver includes those designed for constant amplitude signals, such as FM and many types of PM. Here, in principle, AGC is not required, provided that the IF strip is designed as described in Chapter 6 so that each stage limits cleanly when fed with an input as large as its output. However, in the more sensitive receivers, AGC is often incorporated to prevent overload of the early stages, when for example a car radio passes by an FM transmitter: AGC of the RF stage will prevent mixer overload. Generally one cannot successfully apply AGC to mixers themselves. In addition to AGC, FM receivers will also frequently incorporate AFC (see Chapter 7). There remain two other classes of receivers, both dating from the earliest days of 'wireless': the homodyne and the super-regenerative receiver. The former has in recent years enjoyed renewed popularity, whilst the latter threatens to proliferate also, with possibly unfortunate results.

The homodyne is a single superhet receiver where the LO frequency is equal to that of the carrier of the wanted signal, so that the IF frequency is 0 Hz. One implementation uses an oscillator with a characteristic similar to that in Figure 8.3d as the both the LO and the mixer. The loop gain is adjusted so that the circuit barely oscillates and being very susceptible to outside influences, it is easily tuned so as to become phase locked to the carrier of the incoming signal. This arrangement is also known as a synchrodyne. The modulation of the incoming signal is impressed on the local oscillator and may be recovered with a suitably coupled detector. The upper and lower sidebands of an AM signal are in effect translated down to baseband, and as the oscillator is phaselocked to the carrier (and in phase with it), they lie perfectly on top of each other. The circuit will also receive SSB signals, though in this case there is usually insufficient residual carrier power to take control of the oscillator's frequency, since in SSB the carrier is suppressed by at least 40 dB relative to PEP (peak envelope power). However, as there is only one sideband, the result is quite intelligible provided the mistuning does not exceed about 10 Hz. (Such mistuning on an AM signal would result in one sideband coming out 10 Hz lower in frequency than it should and the other 10 Hz higher, the resulting 20 Hz misalignment garbling the baseband signals.) The homodyne will also receive CW signals, by off-tuning to one side or the other to provide an audible beat. Similarly, it can translate the two tones of an FSK signal to baseband, where they can be picked out by appropriate narrow-band tone filters to recover the message information. However, when using the simple homodyne receiver off-tuned like this to one side of the wanted signal, interference may be experienced from an unwanted signal on the other side of the LO frequency. For an FSK signal, a better approach is to tune the receiver exactly half-way between the two tones, which now appear at baseband indistinguishable as far as their frequency is concerned. However, one is a positive frequency and one is a negative frequency relative to the receiver's LO, and they can thus be distinguished if the sense of their phase rotation is taken into account. To do this, it is necessary to compare the outputs of two homodyne circuits with LO signals in quadrature (Figure 10.3a). Now, if the input frequency is above the LO frequency, the phase of the signal in the upper I (in phase) channel will lag that in the lower Q (quadrature) channel, but it will lead if the input is below the LO. Thus as long as a mark tone persists, a 1 (say) will be clocked into the D flipflop every cycle, and likewise a 0 in the presence of a space tone. The bandwidth of the receiver (which is set by the low-pass filters) need only exceed half the tone separation by a modest margin to allow for the data rate and any possible mistuning, so cut-off frequency of the low-pass filters can be set to say 75% of the tone separation. For even greater selectivity and immunity to interference, band-pass filters could be used. Figure 10.3b shows a complete data receiver suitable for a pocket pager working on this principle: the 90° phase shift between the two local oscillator signals to the mixers is provided by the off-chip 45° lead and lag networks C15,R6 and R7,C13. This system works because in an FSK signal only one tone is present at any one time. Imagine that both tones were present simultaneously and the carrier were 10 Hz off tune. Then each channel would contain both tunes, one 20 Hz higher than the other; the receiver would not be able to separate them. However, now imagine that the baseband signal out of the Q mixer is subsequently passed through a broadband 90° phase shifter. The baseband signal due to the upper sideband will now be in phase in both channels, whilst that due to the lower sideband will be in antiphase. So if the two channels are added, the lower sideband contribution will cancel out leaving only the signal due to the upper sideband, whilst conversely, differencing the I and Q channel will provide just the lower sideband signal. This arrangement is known as an image reject mixer (Figure 10.4). The baseband 90°

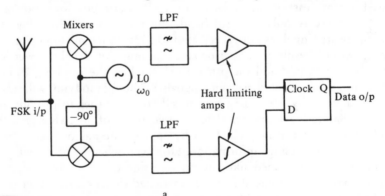

a

Figure 10.3   *Homodyne FSK receivers.*
a Block diagram of a homodyne FSK receiver. (Reproduced by courtesy of Electronics World *and* Wireless World.*)*

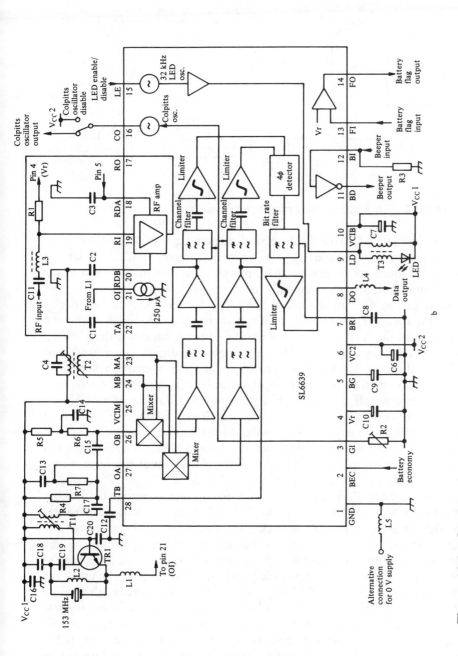

Figure 10.3 Continued
b Complete homodyne FSK receiver circuit. (Reproduced by courtesy of GEC Plessey Semiconductors.)

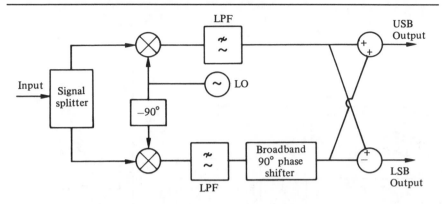

Figure 10.4 *An image-reject mixer (receiving).*

phase-shifter should cover the baseband of interest, e.g. 300–2700 Hz for speech — outside this band the outphasing no longer holds so sideband separation would not be complete. In practice, due to limitations in mixer and channel balance and accuracy of the quadrature phase shifts, the rejection of the unwanted sideband is often limited to about 35–40 dB. Where the image reject mixer is used not at the incoming signal frequency direct, but as the final IF stage in a double or triple superhet, the I and Q signals can be digitized in ADCs (analogue to digital converters) and subsequently corrected for quadrature, gain and offset errors, resulting in greatly enhanced rejection. The homodyne can also be used for the reception of analog FM signals such as NBFM (narrow band FM) voice traffic [1].

The super-regenerative receiver was developed in the early days of wireless to take advantage of the considerable gain in sensitivity which could be achieved by the use of reaction, where a gain of 50 dB in a single stage is possible. With reaction, a proportion of the RF signal at the output of a tuned RF or leaky grid detector stage is fed back to its input. If carried to excess, the stage will oscillate, so it is essential that its characteristic is rather like Figure 8.3d and definitely not like Figure 8.3b. Unfortunately, considerable skill in adjustment was necessary to obtain the full benefit available from reaction, so many listeners could not master the operation. In the super(sonically quenched oscillator)-regenerative receiver, the loop gain of an RF amplifier with feedback is varied cyclically above and below unity at a supersonic rate, typically 100 kHz (Figure 10.5). This is usually achieved by cyclically varying the current drawn by the active device [2]. There is some similarity to the homodyne, but although the sensitivity is increased greatly, the great increase in selectivity achieved with reaction is not obtained. In the absence of any signal from the aerial, the oscillations which build up during each cycle of the quench waveform start from an initial amplitude determined by the noise level in the input circuit and reach an equilibrium value

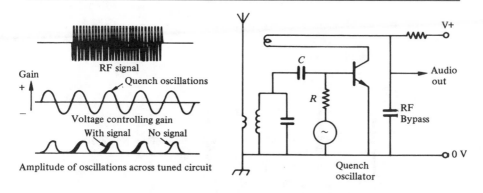

Figure 10.5   *Operation of a super-regenerative receiver.*

equal to the steady oscillation level which would prevail if the circuit were not repeatedly quenched. (This assumes the circuit is being used in the usual 'logarithmic' mode, rather than the alternative linear mode in which the oscillation is quenched before reaching its equilibrium value.) The oscillations die out when the quench voltage reduces the loop gain below unity. For proper operation, the oscillation must decay to a level below circuit noise before the quench waveform again causes the loop gain to exceed unity. If now a signal above noise level is present within the bandwidth of the tuned circuit, when the oscillations start to build up they start from a larger amplitude than before (Figure 10.5). The oscillations therefore reach equilibrium level earlier and the average current drawn by the active device is increased. The signal modulation thus appears as a modulation of the device current, so the device acts as detector as well as amplifier. The equilibrium level of the oscillation and its subsequent decay are not significantly affected by the presence of a signal. A detailed study of this mode of operation reveals that the change in average device current is proportional to the logarithm of the signal amplitude. Thus the reproduction of an AM envelope with a high modulation index is noticeably distorted. However, the logarithmic characteristic exerts a pronounced limiting action, resulting in a much reduced change of output level between large and small signals — a sort of built-in AGC. It also limits the receiver's response to impulsive interference, which in any case is less of a problem than with other types of receiver, since a narrow noise spike will be ignored completely unless it occurs during the brief period of build-up of the oscillation — a small fraction of each quench cycle. The logarithmic characteristic also results in a capture effect, whereby when two signals are present simultaneously, the larger controls the build-up of oscillations, almost completely suppressing the effect of the weaker signal. The circuit of Figure 10.5 shows a separate quench oscillator, but this can often be dispensed with, by making the time constant CR long enough to cause the oscillator to 'squegg'. An oscillator squeggs when

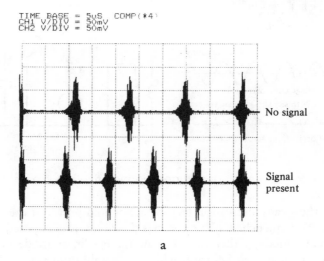

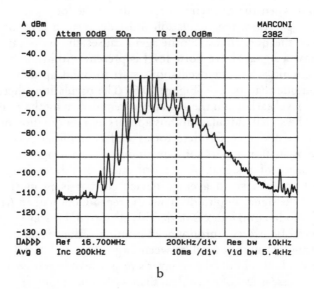

Figure 10.6 *Super-regenerative receiver (self-quenching).*
a Tank circuit waveform.
b Spectrum of a.

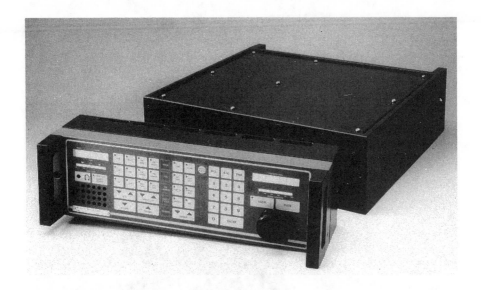

Figure 10.7 *The Skanti R8003 covers the LF, MF and HF bands to 30 MHz with 10 Hz resolution, with scanning and 399 user programmable preset channels. Modes include AM, USB, LSB, ISB, telex and CW. The control unit may be remoted up to 100 m from the receiver unit. (Reproduced by courtesy of Skandinavisk Teleindustry Skanti A/S.)*

operating in a mode where it is self-biasing to class C and the time constant of the self-bias circuit is much too long. The last cycle of the build-up biasses the device back to a point where the loop gain is just less than unity and due to the excessive time constant it cannot recover to unity or above before the next cycle. The oscillation therefore dies away completely leaving the device cut off, until the charge on C leaks away and the device turns on again to the point where the gain exceeds unity. In this self-quenched mode of operation, the quench frequency increases when a signal is present. The information carried by the incoming signal can be recovered from the frequency modulation of the quench frequency, see Figure 10.6a (the individual cycles of RF are not fully delineated by the digital storage oscilloscope used owing to the large difference between the quench frequency and the RF). The super-regenerative system thus offers a simple, compact circuit with high sensitivity at very low cost, which has re-awakened interest in its use at VHF and UHF as a receiver for applications such as remote garage door opening, car central locking, etc. However, if it becomes popular, problems of interference could arise, as it is impossible to design the circuit so that it does not emit energy at the frequency of the oscillator, surrounded by many sidebands at the quench frequency (Figure 10.6b).

a

Figure 10.8
a A modern mobile 'phone operating via the GSM (Global System Mobile) network.
b Block diagram of a Motorola cellular telephone.
(Reproduced by courtesy of Motorola European Cellular Subscriber Division.)

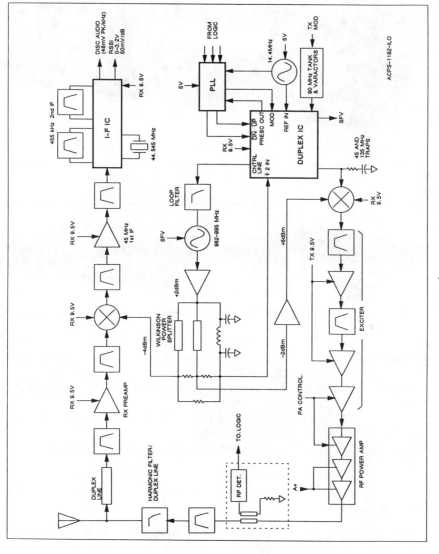

ACPS-1182-ILO

b

a

b

Figure 10.9 *The PTR4300 combat net radio can also interface with many items from the System 4000 common ancillaries subsystem.*
a The PTR4300 ECM-protected combat net radio covers 2–29.9999 MHz at outputs up to 20 W.
b Two of the many optional units from the PTR4300 range, which also includes ATUs, PAs, cositing filters, etc.
(Reproduced by courtesy of Siemens Plessey Defence Systems.)

References

1. Hickman, I. Direct conversion FM design. *Electronics World and Wireless World* November, pp. 962–7 (1990)
2. Terman, F. E. *Electronic and Radio Engineering*, 4th edn, McGraw-Hill, New York, p. 566 (1955)

11
Propagation

This chapter and the succeeding one between them cover the topics of antennas and propagation. Both are very wide ranging subjects, so it will only be possible to scratch the surface in these two chapters. There is a vast quantity of literature relating to each of these topics, and from it, a small selection of references has been included at the end of each chapter, for further reading. In addition to propagation, the topic of external noise (both naturally occurring and man-made) is, for convenience, also covered in this chapter since (together with antenna gains and propagation loss), it determines the transmitter power needed to communicate over any given path.

The topics of antennas and propagation are closely inter-related, so it will be helpful to start a consideration of propagation with a look at the electric and magnetic field distributions both close to and far from a basic dipole antenna, although the main treatment of this antenna is reserved for Chapter 12. Figure 11.1 shows the electric and magnetic fields from a vertically polarized dipole radiator. The electric field is everywhere at right angles to the magnetic field and both are everywhere at right angles to the direction of radiation. (This condition can be met in two dimensions but not in three, which is why an isotropic radiator is not possible. An isotropic radiator would radiate an equal intensity signal — or alternatively receive equally well — in all directions. Although not physically realizable, it is a useful yardstick for comparing other antennas.) The electric lines must start and finish on the conducting elements of the dipole, whilst the magnetic lines must form closed loops encircling the current flowing in those conducting elements. The current flowing in the elements of a resonant $\lambda/2$ dipole is (almost) in quadrature with the applied voltage, so the electric and magnetic fields in space close to the dipole are also in quadrature; this is the 'near field' region. The associated energy circulates back and forth between the electric and magnetic fields, exactly as in a tuned circuit and the Q value of the antenna

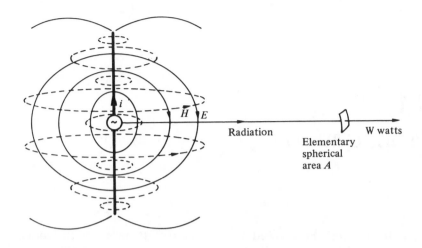

Figure 11.1   *Near and far fields of an antenna.*

determines its 3 dB bandwidth in exactly the same way as for a tuned circuit. When exactly on tune the antenna looks resistive to the source since the latter only supplies the energy 'consumed' by the radiation resistance R_r (and by the loss resistance R_1, although in a well designed efficient antenna, this may amount to as little as a few per cent of the power radiated). The quadrature electric and magnetic fields close to the dipole are called 'induction fields' and they drop off more rapidly with increasing distance from the dipole than do the electric and magnetic components of the radiation field. The latter are in phase with each other and thus describe a flow of power radiating outwards from the antenna.

Beyond a few wavelengths from the antenna, the radiation field greatly exceeds the induction field; this is called the far field region, where the radiated energy expands as a spherical wavefront centred on the radiator. (At a great distance from the antenna, the radius of this spherical wavefront becomes so great, that to a receiving antenna, it appears as a plane wavefront.) The magnetic field is associated with current and the electric field with voltage and their ratio is a resistance. This is called the characteristic resistance of free space, and has the value 120π or 377 Ω. Consider the power W watts flowing through a small area A (in units of square metres) on the surface of such a sphere (Figure 11.1): then the field strength η in volts per metre is given by $\eta = \sqrt{(377\Phi)}$, where Φ is the power density W/A. For each doubling of the distance from the radiator, the power is spread over four times the area. Thus the power available to a receiving antenna falls to one-quarter for a doubling of the distance, giving the attenuation of a radio wave in a lossless medium (free space) is an inverse square law or –6 dB per octave (doubling) of distance. In a radar system, such energy as is scattered by a small target back in the direction of the radar set is also subject to the inverse square law, giving the basic radar range law as R^{-4} or inverse fourth power of range. Where the target fills the field of view of the antenna in one dimension (e.g. the horizon) or two dimensions (large cloud bank), the range law becomes R^{-3} or R^{-2} respectively. By contrast, metal detectors work upon the more rapidly decaying induction field (near field) and so are subject to an R^{-6} range law.

Turning now to a complete radio communication path, the path loss between isotropic antennas in free space, defined as the ratio of transmitted power P_t to received power P_r is $(4\pi d/\lambda)^2$, assuming d (distance) is large compared with λ, d and λ both in metres. For two half-wave dipoles (broadside on to each other), the loss will be less, since each has a gain in the maximum direction of 2.15 dB ($\times 1.65$) relative to isotropic, giving $P_t/P_r = (2.44\pi d/\lambda)^2$; so for example at a spacing of 10λ, the received power is 1/5876 times the transmitted power. Due to the –6 dB/octave (inverse square) law, the received power will be four times as great every time d is halved. On this basis, when the separation is $1/(2.44\pi)$ times a wavelength, there is no loss at all between a pair of half-wave dipoles, and at half this separation the received power is four times as great as the transmitted power! Of course, the formula only holds for the far-field region, not for a spacing as small as $\lambda/(2.44\pi) = 0.13\lambda$. Nevertheless, using 0.13λ as a starting point, with a little practice at the mental arithmetic you can astound your colleagues by working out the free-space path loss for a communications system in your head. For example, at 144 MHz λ is approximately 2 m and at a separation of 0.25 km (approx. 1000 times 0.13λ or 2^{10} times or 10 octaves of distance), the free-space loss between half-wave dipoles is simply $(10 \times 6) = 60$ dB. An alternative starting point that can be useful to memorize, is that the path loss between isotropic antennas separated by a distance equal to λ, is 22 dB.

Where the antennas have a different value of gain, this must be allowed for, leading to the formula

$$P_t/P_r = (4\pi d/\lambda)^2/(G_t G_r)$$

where $G_t G_r$ is the power gain relative to isotropic of the transmit, receive antenna in the required direction respectively.

The above formula may be re-expressed to give the free-space path loss L in decibels as follows

$L = (32.44 + 20 \log_{10} f + 20 \log_{10} d)$ dB, for the case of isotropic antennas
    ($G_t = G_r$ = unity), or
$L = (28.15 + 20 \log_{10} f + 20 \log_{10} d)$ dB, between half-wave dipoles
    ($G_t = G_r = \times 1.65$),
where frequency f is in MHz and distance d is in km.

In many cases we need to know the path loss taking into account the effect of the surrounding terrain. The following deals only with paths short enough to be considered as over flat earth; for paths long enough for the effect of the earth's curvature to be important, the range is generally determined by factors other than those considered below. The following also refers to cases where the ground wave can be neglected, namely higher frequencies: ground wave propagation is dealt with in a later section.

Figure 11.2a shows antennas that are vertically polarized, but the following applies also to horizontally polarized antennas. The voltage induced in the receiving antenna is the resultant obtained by adding the direct and the reflected rays. If the angle θ at which the incident ray strikes the ground is very small, then the reflected ray will suffer a phase reversal. In the case of smooth ground (or calm water), the reflected ray is little attenuated (even if the ground is of poor conductivity) and so its magnitude at the receiving antenna will be nearly the same as the direct ray. If the difference in the lengths of the paths taken by the direct and indirect rays is small compared with the signal's wavelength λ, then the two versions of the received signal will be nearly in antiphase. Under these conditions, the received signal amplitude will be directly proportional to the phase shift between the two rays, Figure 11.2b. The received signal level will therefore be considerably less than it would be if the direct ray

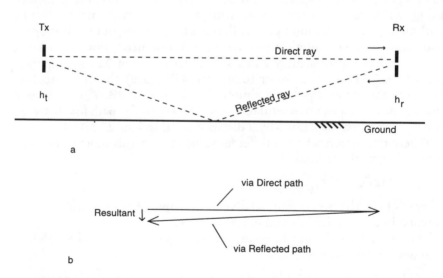

Figure 11.2
  a Propagation over a flat earth path.
  b Showing how the net received signal is much lower than would be the case for a path in free space.

were received in the absence of the reflected ray. From the geometry of the situation and taking account of both the free-space loss and the additional loss due to cancellation, the ratio of received to transmitted power P_r/P_t between isotropic antennas mounted at heights h_t and h_r separated by distance d is equal to $(h_t h_r/d^2)^2$, independent of units, provided both height and distance are in the same units, e.g. metres.

Note that unlike the free-space loss, this does not increase with frequency since as λ gets shorter, the phase shift between the direct and incident rays increases and hence so does the resultant. Note also that if the range is doubled, the antenna heights remaining unchanged, then due to the geometry (the angle between the direct and the reflected ray being halved) the angle between the vectors representing the direct and reflected rays in Figure 11.2b will also be halved. Thus the size of the resultant *relative to the direct ray* will be halved. But the direct (and reflected) ray is itself halved in amplitude, due to the doubled range. Thus the path loss is now proportional to the fourth power of d, i.e. the range law is now −12 dB/octave of distance. Be careful when using this formula; remember it only applies if the phase shift between incident and reflected rays at the receive antenna is small. Always work out the free-space loss as well and distrust the original answer if it is not much greater than the free-space loss.

Both the free space and flat earth formulae above assume straight ray (LOS — line-of-sight) propagation. This is not always the case. Where a LOS path does not exist, communication may still be possible. In this case, the signal reaches the receiver by diffraction, or by penetration (more effective at lower frequencies), or by reflection (more effective at higher frequencies). For communication to be successful, the additional losses must be allowed for. These can be calculated for simple cases, or use may be made of measured values published in the literature. A great deal of work has been done on propagation at VHF and UHF in connection with PMR (private mobile radio) and mobile telephones, e.g. [1]. In this case, the base station antenna is elevated, but the mobile's antenna is not, and will frequenctly be screened. A well known study was carried out by Egli (one of the earlier workers in the field) [2]. From a study of a large number of measurements made in large towns, he suggests that at frequencies above 40 MHz, an additional empirically-derived term $(40/f)^2$ (f in MHz) be inserted in the above equation. This is a median allowance for base-to-vehicle and vehicle-to-base paths: he also gives statistical spreads, which differ for the two cases. Figure 11.3 shows the predicted path loss versus range for comunications in the region of 140 MHz.

The flat earth propagation formula, together with empirical adjustments suggested by Egli, Okamura and others, give good guidance to the maximum range which can be expected for a given transmitted power

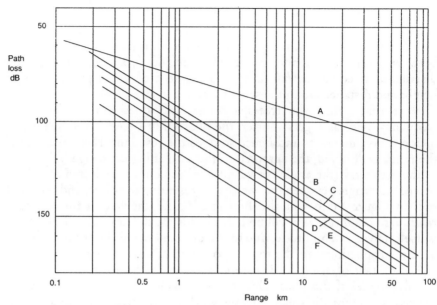

Path loss versus range at 140MHz. A: Free Space. B: EGLI 50%. C: CCIR 50%.
D: EGLI 90%.   E: CCIR 90%.   F: OKUMURA 50% (URBAN)

Figure 11.3 *Predicted typical path loss for communications at 140 MHz. The 12 dB/octave of distance contrasts with the 6 dB/octave of propagation in free space. There is a difference of just over 4 dB between the Egli and CCIR figures. This could be because the former are possibly given for loss between dipoles, the latter between isotropic antennas.*

at VHF and above, at least out to the 'radio horizon'. The factors determining the distance of the radio horizon are complex, including antenna heights among other things. But briefly, the radio horizon is the distance beyond which the received signal strength falls off very rapidly. So rapidly in fact, that there is an upper limit to the transmitter power that it is worth using with a given antenna height. However, VHF/UHF signals may occasionally be received at distances well beyond the radio horizon, due to conditions such as a temperature inversion, ducting, etc., the effects often being evident as, for example, patterning on a TV set.

At HF and lower frequencies (30 MHz downwards) the same formulae still indeed apply, but the actual range is often found to far exceed that thus predicted for various reasons. Firstly, at lower frequencies, radiated power travelling parallel to the earth is slowed down at the earth/air interface due to the conductivity and the high dielectric constant of soil or water. As a result, the wavefront instead of being vertical, tends to tilt forward at higher levels and thus to follow round the curvature of the earth: this is known as the ground wave. Note that the ground wave is

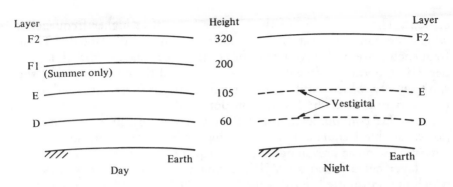

Figure 11.4 *Ionosphere: heights of layers in kilometres (approximately)*

always vertically polarized; the conductivity of the earth short circuits any horizontally polarized component of the wave, eliminating any horizontal component of electric flux. At low frequencies the ground wave range is very extensive, so that for instance the BBC's Droitwich transmitter (whose 198 kHz carrier frequency is maintained to an accuracy of 1 part in 10^{11}) can be received over much of continental Europe.

At even lower frequencies such as VLF (very low frequencies, 3–30 kHz) the ground wave extends for thousands of kilometres (an earth-ionospheric waveguide duct mode is also relevant here) and even penetrates the surface of the ocean very slightly, so that VLF can be used for world-wide communication with submarines, albeit at a very restricted data rate. At HF, the ground wave falls off much sooner: nevertheless long distance communication is still often possible. This is because ionized layers of the atmosphere (the 'ionosphere') reflect back towards the earth signals that would otherwise be lost into space (Figure 11.4). The signal, on striking the earth, is reflected and may then be reflected from the ionosphere a second time, to return to earth even further away. The distance from the transmitter to where the first reflection strikes the earth is known as the 'skip distance' and the area of no reception beyond ground wave range to where the first reflected signal is heard is known as the 'dead zone'.

During the daytime, typically there are four ionized layers at different heights. The lowest, the D layer, is responsible for heavy attenuation at MW frequencies, giving interference-free reception of MW broadcast stations within their ground wave range during the hours of daylight. After dark, it almost disappears as in the absence of sunlight, the ions and electrons recombine; distant MW stations can then be heard via ionospheric reflections at ranges way beyond their intended primary ground wave service area, leading to severe interference with local stations. The attenuation of the D layer falls off at higher frequencies, which can thus penetrate it even during the hours of daylight. These frequencies are reflected from the E layer or one of the F layers, depending upon the time of day, the season and the current level of the sun's activity, which exhibits short-term variations (over days) and long-term variations over the 11-year sunspot cycle.

For an HF communications link there will be at any given time an LUF (lowest usable frequency) set by the higher levels of absorption and of atmospheric noise prevailing at lower frequencies, and other factors such as E-layer cut-off, and an MUF (maximum usable frequency) beyond which the transmitted signal penetrates all the layers and does not return to earth. The strongest return occurs at just below the MUF, but it is better to work at a slighty lower frequency to allow for slight short-term variations in the MUF. Typically, communication is carried out at a frequency of about 85% of the monthly median of the F2 MUF; this is known as the OWF (optimum working frequency) or the FOT (frequence optimum de transmission) and is assumed to give a path for about 90% of the time, assuming communication is possible. For it can happen occasionally that the frequency range between the LUF and the MUF becomes vanishingly small. Though not a common occurrence, this is most likely to occur on long paths where part of the path is in daylight and part in darkness, or in trans-polar paths where high levels of absorption may raise the LUF until it equals or exceeds the MUF.

Choice of operating frequency may be left to the judgement of an experienced operator, choosing from among a limited number of assigned frequencies. However, experienced operators are becoming rare whilst the demand is for ever more reliable HF communications. To this end, computer programs are available to assist in calculating the best operating frequency for any given route at any given time; this might be for example a three-hop path via the E layer (3E) and/or a one-hop path via the F2 layer (1F2). Examples of such programs are APPLAB 4, from the Rutherford Appleton Laboratory, Didcot, Oxfordshire, UK, and 'Muffy'. The latter program, though less sophisticated, can be run on a PC or compatible personal computer and is thus popular with amateurs.

Typically, a prediction program will give the required transmitted power for any paths that are 'open', taking into account the latitude and longitude of the transmitter and of the receiver and their heights above sea level, the receiver bandwidth, the type of antenna, the time of day, season, and sunspot number. Propagation prediction programmes can only take into account known average conditions; they are unaware of any incidental short-term variations from these mean conditions. In particular, it would be wrong to think of the various ionized layers as perfect spherical mirrors encompassing the globe. In places they may exhibit dents, corrugations or other irregularities. These are transitory disturbances due to wind shear and other meteorological effects, with the result that a path between a transmitter and a receiver, predicted as open at a certain frequency by a program such as Applab, may in fact not be available to pass traffic, whilst a path not predicted as open may well provide an excellent signal at the receiver. There are also other more catastrophic effects, all associated with solar flares, traditionally considered unforecastable though hopefuly progress is being made in this direction. These effects include:

- Sudden ionospheric disturbances (SIDs): caused by UV and X-rays; greatly increased D layer absorption plus other effects; follows closely on flare; usually lasts from a few minutes to a few hours.
- Ionospheric storms: caused by protons and electrons; depression of F2 critical frequencies plus other effects; 20–40 hours after the flare; can last for up to 5 days.
- Polar cap absorption: caused by protons; high absorption; a few hours after the flare; lasting 1–10 days.

It will be apparent from the foregoing that a certain amount of uncertainty exists as to whether communication is possible over a given path on one of the assigned frequencies available to the would-be communicator. Consequently, use may be made of another advanced aid to HF communications reliability, namely the chirp sounder. Various stations around the world transmit at different times at precisely known intervals a CW transmission which sweeps steadily across the whole HF band. A special purpose chirp receiver can receive the signal from the chirp sounding transmitter, displaying received signal strength and time delay of the signal versus frequency. The former enables a frequency offering an adequate signal to noise ratio to be chosen whilst the latter permits the avoidance of frequencies at which two or more paths are open. This is particularly beneficial for radio-telex or data transmissions, to minimize errors due to ISI (intersymbol interference). The time delay difference between paths is typically 2–3 ms with a normal maximum of 5 ms and a worst case of about 10 ms. Interestingly, the largest spread of delays is in fact experienced over short paths.

Where a special purpose chirp receiver is not available, use can still be made of chirp transmissions. It is only necessary to listen out on the intended frequency of communication (or an adjacent clear channel) for a chirp transmission from a transmitter near to the other end of the intended link. A characteristic up-chirp will be heard (or a down-chirp if using lower sideband) as the transmission sweeps through the receiver channel. Knowing the expected time of the sweep passing through the tuned frequency, and given an accurate clock, reception of the chirp will indicate that the path is open. By listening on other frequencies, the current values of the LUF and MUF, for the given path, can be estimated.

Chirp-sounding transmitters are operated at various sites in the UK by various branches of the services, and by certain other agencies throughout the world at sites ranging from Oslo (NATO), Belize, Norfolk Virginia, the Phillipines, Hong Kong, Canada, Saudi Arabia (with no less than three transmitter sites) and others. All stations transmit at the same sweep rate of 10 seconds per MHz, thus taking 4 minutes 40 seconds to cover the band 2–30 MHz. Some stations transmit a chirp every 15 minutes, others every 5 minutes. Each station has a unique start delay of so many minutes and seconds past the hour (or past the quarter hour, etc.), so that knowing this, and given the 10s/MHz sweep rate, the exact expected chirp time for any given transmitter can be determined for any particular receive frequency. Thus, given an accurate watch, any chirp received indicates an open path to the general location of the corresponding chirp transmitter.

The three ionospheric effects listed above and other variations also have an effect upon DF (direction finding) systems. SITs (systematic ionospheric electron density tilts) may result in an HF signal returning to earth at a different point from where it would have appeared had the ionosphere been smooth and regular. This can introduce an error in the measured bearing of the transmitter at one or both receiving stations of a DF system, resulting in the position indicated by the intersection of the cross bearings being inaccurate. SITs [3] have a particularly serious effect on single stations DF systems, which rely on measurement of the azimuth and elevation arrival angles, and an estimate of the height of the appropriate reflecting layer, to calculate both the bearing and distance of the target transmitter. Similarly, TIDs (travelling ionospheric disturbances) [4] produce gradients in the electron density, again resulting in propagation of an HF signal over a path which deviates from a great-circle direction.

Transmissions at frequencies above about 28 MHz normally pass through all the layers and do not return to earth. However, they may still be used for over-the-horizon communications in certain circumstances. A troposcatter link operates at microwave, depending upon irregularities in the troposphere to scatter a highly directional beam of microwave energy transmitted at a low elevation angle. Sufficient energy is directed back down again in a forward direction to permit reception at distances well beyond the horizon. There is also ionospheric scatter, which depends upon irregularities in the D layer. Meteorscatter communications use frequencies in the range 35–75 MHz. Here, communication is by reflection from the trail of ionized air left by the passage of a meteorite. This acts as a 'wire in the sky', capable of reflecting the incident energy to the receiving end of the link, if the polarization and orientation are right. The transmitting station repeatedly sends a short 'message-waiting' transmission, and on receiving a reply from the intended recipient, sends text, a packet of data or other message as required. The geometry of the path is critical, so that it is unlikely that the signal can be intercepted by other than the intended receiving station. As with troposcatter, for a fixed link, directional antennas can be employed with advantage. The abundance of meteor trails depends upon the time of day, season and latitude, so the waiting time for a path to occur may be anything from a few seconds to many minutes. The length of time for which a trail persists is anything from a few tens of milliseconds to a few seconds and during this time it offers a high integrity path capable of supporting a data rate of up to 10 kb/s or more. The unpredictable waiting time makes meteorscatter unsuitable for real-time traffic, but it is ideal for store-and-forward message operation.

In any radio communications link, noise at the receiver sets the lower limit of signal strength which provides a usable signal. A received SNR (signal to noise ratio) of about +10 dB is required for speech and a similar figure suffices for fairly robust forms of digital modulation. The most robust types can operate with a signal to noise ratio of 0 dB or even a small negative SNR, as can a good CW morse operator, whereas very bandwidth-economical methods of modulation such as 64QAM or 256QAM (carrying 6 bits or 8 bits per symbol respectively) require a signal to noise ratio in excess of 20 dB. By contrast, a 'direct sequence' spread spectrum system (where the actual data rate is much lower than the modulation or 'chipping' rate), can provide up to 25 dB or more of 'processing gain', permitting such a system to operate with a large negative signal to noise ratio.

The noise at the receiver comes from several sources. The first is the receiver's own noise (internal noise), mainly attributable to the first active stage such as RF stage or first mixer; this noise is considered in earlier chapters. The noise with which we are concerned here is external noise and this arises from three sources. Atmospheric noise is mainly due to electrical storms in the tropical regions of the world, although other sources such as the *aurora borealis* (Northern Lights) and the *aurora australis* also contribute. The intensity of atmospheric noise varies with the time of day, season and the 11-year sunspot cycle, and also the geographical location of the receiver.

The second type of noise is galactic noise, which is of cosmic origin. This is largely invariant in intensity which is greatest in the direction of the galactic centre; it is only of importance in the frequency range 3–300 MHz, and then only at times and seasons of low atmospheric noise, and at sites where man-made noise is low.

The third and in many cases the most important type of noise is man-made noise. This arises unintentionally from a wide variety of sources and is either impulsive, e.g. from electric motors, vehicle ignition systems, light switches, thermostats, etc., or continuous such as radiation of clock frequency harmonics from computers, radiation from ISM (industrial, scientific and medical) RF generators used for diathermy, metal treatments, polythene sealing, etc. Man-made noise does not include disruption of radio reception by other radio transmissions (interference) — although in practice this may often be the major problem — or by deliberate attempts to prevent communication (jamming).

The levels of atmospheric noise experienced at various locations throughout the world at various times of day, season and phase of the sunspot cycle are comprehensively listed in Reference 5. Atmospheric noise usually predominates at frequencies up to 30 MHz and the report consequently concentrates on this frequency range. It should be noted that when a directional HF antenna located in temperate latitudes is used, the level of atmospheric noise encountered will be greater if the main lobe points towards the tropics than if it points towards the pole. At frequencies in excess of 100 MHz a receiver is likely to be internally noise limited. (However, note that at any frequency, an inefficient antenna, antenna feeder loss and the insertion loss of any filters ahead of the first stage of amplification will all attenuate both the wanted signal and the external noise, possibly leading to the receiving system being internally noise limited.) At microwave frequencies the external noise level is so low that (unless the antenna is pointed at a noise source, e.g. the sun) for very weak signals it is useful to take steps to reduce the receiver's noise figure below the thermal noise level prevailing at room temperature. This may be done either by refrigerating the RF amplifier in liquid nitrogen or liquid helium, or by using a parametric amplifier. When designing a receiver it is useful to have guidance as to the *minimum* likely level of external noise, since there is no point in incurring additional cost to secure a receiver internal noise level much lower than this. Reference 6 gives this information for frequencies from 0.1 Hz to 100 GHz, covering atmospheric, galactic and man-made noise. For much of this frequency range it also gives some useful guidance as to the likely maximum levels. Figures 2 and 3 from this report are reproduced in this volume, by permission of the ITU-R, as Apendix 12. Between them, they more than cover all the frequencies used for radio communication with which this book is concerned, i.e. principally from 100 kHz to 1000 MHz.

References

1.   Ibrahim and Parsons. Urban mobile radio propagation at 900 MHz. *Electronics Letters*, **18**(3), 113–15 (4 February 1982)
2.   Egli. Radio propagation above 40MC over irregular terrain. Proceedings of the I.R.E., pp. 1383–91 (October 1957)
3.   Tedd, Strangeways and Jones. Systematic ionospheric electron density tilts (SITs) at mid-latitudes and their associated HF bearing errors.

Journal of Atmospheric and Terrestrial Physics, **47**(11), 1085–97, 1985
4. Tedd, Strangeways and Jones. The influence of large scale TIDs on the bearings of geographically spaced HF transmissions. *Journal of Atmospheric and Terrestrial Physics*, **46**(2), 109–17, 1984
5. International Telecommunication Union, *World Distribution and Characteristics of Atmospheric Radio Noise*, CCIR Report 322, Geneva (1964)
6. International Telecommunication Union, *Worldwide Minimum External Noise Levels, 0.1 Hz to 100 GHz*, CCIR Report 670, Geneva (1978)

12
Antennas

An antenna is a device designed to accept RF power from a transmitter and radiate it into its surroundings, or alternatively to extract energy from a passing radio wave and deliver it to a receiver. Considering transmitting first, an antenna is ideally designed to present a resistive load $R_t = R_r + R_l$ (a pure resistance equal to the design load impedance of the transmitter, usually 50 Ω, if perfectly tuned and matched) and it is to this resistance that the transmitter delivers power. If the antenna is also loss-free, all the power delivered to it goes into the radiation resistance R_r and is radiated; if not, a proportion of it is converted into heat in the antenna's loss resistance R_l. The efficiency η of an antenna is given by $\eta = R_r/R_t$. An ideal isotropic antenna is loss-free and radiates power in all directions with an equal intensity; it is a figment of the imagination as Maxwell's equations describing electromagnetic radiation do not permit of such a design, but it is a useful yardstick for practical antennas.

Practical antennas fall into two main groups, those which are self-resonant and those which are not. But note that in use, non-resonant antennas are often brought to resonance, e.g. with the aid of an ATU (antenna tuning unit; see Figure 12.8. The simplest resonant antenna is the half-wave dipole (known in the Americas as a doublet), the fields in the vicinity of which are shown in Figure 11.1. Figure 12.1a shows its figure-of-eight vertical radiation pattern in cross-section. The radiation intensity is a maximum in the plane at right angles to the dipole and is 'doughnut' shaped; there is no radiation along the line of the dipole. A vertical dipole is described as 'vertically polarized' since the lines of electric field in the direction of maximum radiation are vertical. As can be seen, the two halves of the figure eight are not quite circular. They are exactly circular for a dipole very much shorter than half a wavelength, but such an antenna is not resonant. In the direction of maximum radiation, the field strength produced by a lossless resonant $\lambda/2$ dipole is 1.28 times that of an isotropic radiator, or '2.15 dB above isotropic', whilst for a

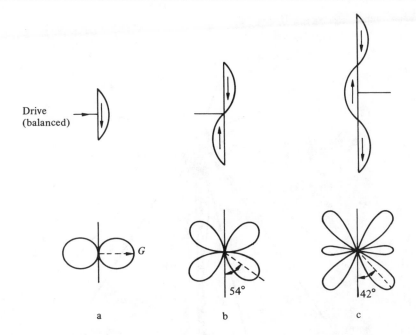

Figure 12.1   *Current distributions on, and vertical radiation patterns of, vertical dipoles remote from the ground. The power gain G of an ideal lossless λ/2 dipole in horizontal plane is G = 1.65 (+2.15 dB) relative to isotropic.*
 a Length = $\frac{1}{2}\lambda$.
 b Length = λ.
 c Length = $\frac{3}{2}\lambda$.

(suitably-matched loss-free) short dipole it is 1.22 times (1.76 dB). When considering a *perfectly-matched lossless* dipole, these figures also represent the 'directivity' or gain relative to an ideal isotropic antenna. However, the term 'gain' should be restricted to the ratio of the *actual* maximum field produced by an antenna, relative to that which would be produced by an ideal isotropic antenna, i.e. 'gain' takes into account an antenna's losses due to R_1. Only in the case of a perfectly-matched lossless antenna does the directivity equal the gain in the maximum direction. In the case particularly of antennas which are not self-resonant, the difference between gain and directivity can sometimes be very large, even when the antenna is brought to resonance by tuning.

Due to end effects, a thin wire radiator such as that in Figure 12.1a has an electrical length which is about 0.025λ longer than its physical length. Like all resonant circuits, a resonant antenna has a bandwidth depending upon the circuit constants. For thin wire dipoles — length/diameter of the

Figure 12.2   *HC30-76 'helicone' skeleton discone antenna, rated 30–76 MHz, 50 W. The elements are plastic-sheathed copper-plated steel helical springs so the antenna is small, light and virtually unbreakable. (Reproduced by courtesy of Racal Antennas Ltd.)*

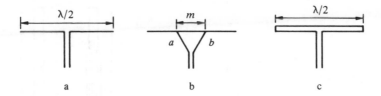

Figure 12.3   *Half-wave dipoles: feed methods.*
a Centre-fed antenna.
b Tapped antenna.
c Folded dipole.

order 500:1 — the useful bandwidth for transmitting is about +/–10%, limited by the increase in VSWR away from the resonant frequency; rather more for receiving, where a worse VSWR is usually acceptable.

The bandwidth of a dipole can be increased by making the conductors very fat — tubes or wire cages — over most of their length, tapering conically to the feedpoint. A variant on this theme, the discone antenna, is illustrated in Figure 12.2. The operating frequency range may be increased if an ATU (antenna tuning unit) is used to bring the dipole back to resonance. The ATU actually decreases the 'instantaneous bandwidth', but the ATU can retune the dipole to resonance when a different operating frequency is required. For very broadband signals, the instantaneous bandwidth of an antenna can be increased by a technique known as compensation [1]. The impedance of a centre-fed $\lambda/2$ dipole (Figure 12.3a) is low and resistive, typically 73 Ω balanced. To generalize, it is low for dipoles an odd number of half-wavelengths long, and high for an even number of half-wavelengths (e.g. Figure 12.1c and b respectively) as is clear from the current and voltage distributions. For other lengths the impedance is not resistive; such dipoles are not resonant. The radiation patterns for dipoles having lengths of multiples of the half-wavelength at the operating frequency show additional lobes, e.g. for lengths 1 and 1.5 times the wavelength (see Figure 12.1). Note that the number of lobes is equal to twice the number of half-wavelengths. The patterns shown are for antennas in free space, i.e. remote from the ground, which would act as a reflector and modify the patterns.

The 73 Ω impedance of the half-wave antenna of Figure 12.3a is not convenient for connecting to a balanced twin wire feeder, which usually has an impedance of about 300 Ω, but this can be accommodated with a 'delta match' (Figure 12.3b). On the other hand, 75 Ω coaxial cable is about the right impedance for direct connection, but is unbalanced. A 1:1 ratio balun transformer (see Chapter 3) could be used, but this is a broadband device which is rather a waste as the dipole is inherently a narrow band

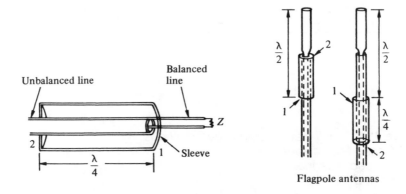

Figure 12.4   *Matching balanced antennas to unbalanced feeders.*
  a Sleeve balun (sleeve shown sectioned).
  b Dipole driven by (unbalanced) coax. The outer-to-sleeve shorts at 2 reflect an open circuit (sleeve to outer) at 1.
  c Alternative construction.

radiator. A narrow band balun can be realized in various ways as in Figure 12.4, and with proper choice of dimensions can also match the antenna to a 50 Ω cable, this impedance being preferred for transmitting systems. For receiving, e.g. for UHF Band IV/V TV, 75 Ω coax is commonly used without a balun, the balanced to unbalanced transition taking place gradually over a distance of several wavelengths along the feeder. Note that a wavelength in the cable is only about 0.7λ, as the velocity of the signal in the cable is only about 70% of that in free space. For VHF FM, a balanced 300 Ω twin wire feeder is often used and here the folded dipole of Figure 12.3c is useful. The two close-spaced dipoles act as a 2:1 turns ratio transformer, transforming the 73 Ω impedance of the simple λ/2 dipole to 292 Ω. A feeder which passes close to a source of interference is less prone to pick-up if it is balanced; in the case of an unbalanced feeder, an interference voltage may be induced in series with the outer, dividing (not necesarily equally) between the antenna and the receiver. In the case of a balanced feeder, the interfering voltage is induced equally in both conductors of the pair as a common-mode or 'push–push' signal, whereas the receiver (ideally) only responds to the normal mode (transverse or push–pull) voltage between the conductors. Incidentally, a folded dipole is often used in a Yagi multi-element antenna, connected to a 75 Ω feeder. The explanation is that one effect of the parasitic elements (reflector and directors) is to greatly reduce the impedance of a simple λ/2 dipole: using a folded dipole restores the desired 75 Ω impedance level.

The antennas which have been considered so far are balanced types. The operation of unbalanced antennas can be approached by looking at the

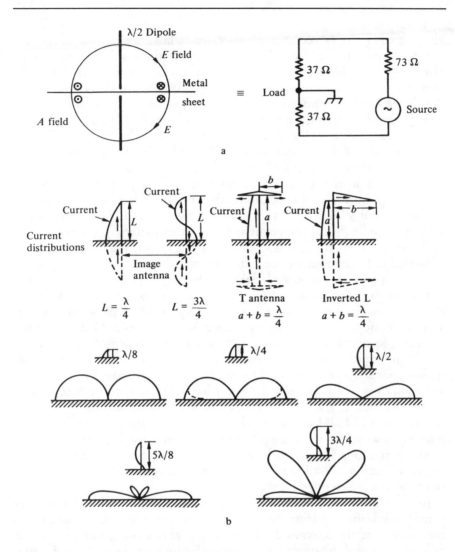

Figure 12.5 *Monopole antennas are unbalanced radiators.*
a Quarter wave groundplane monopole derived from halfwave dipole.
b Current distributions radiation patterns (vertical plane) for various vertical monopoles. All are omni-directional in the horizontal plane.

performance of a modified balanced antenna. Figure 12.5a shows a vertical $\lambda/2$ dipole with a horizontal metal sheet of very high conductivity and infinite in extent (a copper sheet extending many wavelengths would be an adequate approximation) inserted between the two halves, and its equivalent circuit. Note that the electric lines of force all meet the metal sheet at right angles and so are unaffected, whilst the circular horizontal magnetic lines, being parallel to it, do not cut the conductor and so are also unaffected. Therefore the field pattern is likewise unaffected, half the power being radiated above the plane and half below. If now the lower dipole element is removed and all the power fed into the top element (taking care to match the altered input impedance of 37 Ω, the far field of a $\lambda/4$ monopole above a conducting plane is seen to be the same shape as the upper half of the pattern for a $\lambda/2$ dipole but 3 dB higher in strength, or 5.15 dB above isotropic. The conducting plane is usually a 'ground plane', e.g. soil of very good conductivity. If the ground plane is not perfect (e.g. normal soil conditions) then the main lobe does not extend down to ground level. This is shown dotted in Figure 12.5b for the case of a $\lambda/4$ monopole (but applies equally to the other patterns), and the VSWR of the antenna will be high. The VSWR can be greatly improved with a set of buried radial conductors or a chicken-wire earth mat extending out to a radius equal to the antenna height, but for any significant improvement in the low angle radiation the mat would need to extend so much further that it is usually not economic so to do. Figure 12.5b shows the case of various monopoles including top loaded $\lambda/4$ monopoles (T and inverted L, useful to minimize antenna height when the wavelength is long), and the $\frac{3}{4}\lambda$ monopole. Monopoles up to $\lambda/2$ high have only the main lobe, which comes down to ground level; at $\frac{5}{8}\lambda$ small secondary lobes appear and at $\frac{3}{4}\lambda$ these are as large as the lower lobes. (Note that the descriptions T and inverted L are usually applied to antennas which are very much shorter than $\lambda/4$ and consequently not self-resonant even with the top loading, and must be brought into resonance by inductive loading. Medium and long wave broadcast antennas are of this type. Here, the top capacity loading is used to bring the effective height of the antenna closer to the physical height.)

In the case of an antenna elevated above ground, the situation is more complicated, the radiation pattern in the vertical plane depending upon the pattern of the antenna itself, its height above the ground plane, its polarization, and the nature of the ground. Horizontally polarized waves suffer a phase reversal on reflection, exactly so and without loss if over a perfect ground plane. Thus there may be considered to be an 'image' antenna below ground, energized in antiphase. Since all points at ground level are equidistant from the antenna and its image, there is no net radiation at zero elevation. Vertically polarized waves are not phase reversed at angles above the 'peudo-Brewster angle', but are phase reversed below it. For perfect ground, this angle is zero, giving a maximum of radiation at zero elevation angle. But in practice, with normal or even 'good' ground, the peudo-Brewster angle is not zero, so that for rays at grazing incidence, there is phase reversal on reflection and hence a null at zero elevation.

In the case of either horizontally- or vertically-polarized antennas, the radiation pattern in elevation may exhibit one or more lobes, depending upon the antenna height above ground. The greater the height (in wavelengths) of the antenna above ground, the more lobes will appear. On the other hand, the horizontal plane or azimuth pattern depends upon that of the antenna itself, so for the vertical dipole it will be omnidirectional, and for the horizontal dipole basically figure-of-eight.

Many investigations of the radiation patterns of various antennas have been carried out, both in simulation and by actual measurements (e.g. by overflying by helicopter fitted with a measuring antenna). Given the many different types of antenna, varying mounting heights and allowing for the wide range of frequencies used for communications, the possible permutations are infinite. Figure 12.6 shows a computer-simulated radiation pattern of a horizontal half-wave dipole for use at 14 MHz, mounted at a height of $\lambda/2$ (10.7 m) above varying types of ground. The plot shows the radiation pattern in elevation, for a bearing of zero degrees in azimuth, where the radiation is a maximum, i.e. at right-angles to the line of the dipole. It can be seen that the size and shape of the main lobe is little affected by the ground conditions, whereas these strongly affect the radiation in the vertical direction, for the following reason.

Given the stated mounting height, the downward radiation reaches the ground in antiphase. Upon reflection, it suffers a phase reversal, so that the reflected wave *at ground level* is in phase with the upward radiation at the antenna itself. But by the time the reflected wave arrives back at the antenna, it is again in antiphase with the upward radiation at that point and therefore tends to cancel it. If the terrain beneath the antenna is a very good reflector, the reflected wave is barely reduced in amplitude, and so the cancellation is almost complete. Over poor ground, some of the energy radiated downwards penetrates the ground and is absorbed, whilst what is reflected may suffer a phase 'reversal' which is not exactly 180°. Thus the reflected wave arriving back at the antenna is reduced and cancellation is incomplete, leaving appreciable net radiation in the vertical direction.

If the antenna height is raised, the null (or minimum) in the vertical direction splits into two, either side of the vertical. The angular spacing between them increases as the height is raised further, with further nulls successively appearing and splitting likewise.

At a higher frequency, e.g. 30 MHz, the vertical radiation pattern of a horizontal half-wave dipole mounted at the same height in terms of λ, namely $\lambda/2$, is very similar to that of Figure 12.6. But mounted at the same *physical* height as the antenna of Figure 12.6, namely 10.7 m or approximately one wavelength, there will be two distinct lobes either side of the vertical. There is a deep null between them at an elevation angle of about 45°, where the radiation is 8 dB or more below isotropic in the case of

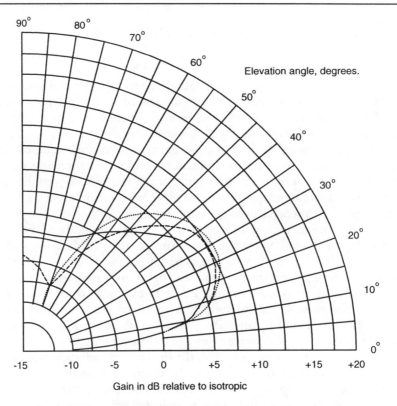

Elevation angle, degrees.

Gain in dB relative to isotropic

Radiation pattern of a 14MHz horizontal half wave dipole (in a vertical plane at right angles to the dipole) mounted at a height of half a wavelength, over the following types of terrain:-

SOIL   -------- (GOOD)   SEA-WATER ....
              ———— (POOR)

Figure 12.6 *Radiation pattern in a vertical plane at right angles to a 14 MHz horizontal half wave dipole, mounted at a height of λ/2 (10.7 m), over various types of terrain.*

good ground (high conductivity and permittivity) — much more in the case of sea water. On the other hand, with poor soil the null is only some 5 dB below isotropic, clearly better if the only path open to a distant receiver involves a take-off angle of 45°. Thus 'good' soil is not necessarily an advantage. With a 30 MHz half-wave horizontal dipole mounted at a height of 2λ m there are four lobes either side of the vertical. The deepest of these, at an elevation angle of about 14°, is very deep regardless of soil type, being some 10 dB below isotropic, the higher nulls being progressively less deep, except in the case of sea water. In many cases, HF communications are typically required over paths of a given length; mainly short paths — for example tactical comms — or alternatively mainly medium to long paths, e.g. diplomatic traffic. Thus an antena mounting height would be chosen to avoid a null at the required take-off angle over the usual range of operating frequencies.

An antenna is a reciprocal device, exhibiting the same polar pattern when receiving as when transmitting. However, when transmitting, the surrounding field is a spherically expanding wavefront centred on the antenna. As a receiver, the antenna experiences a passing plane wavefront, which excites an emf at the antenna's terminals. For a $\lambda/2$ dipole, the emf is $2/\pi$ times lE, where l is the length of the dipole in metres and E is the field strength in volts per metre. The emf is in series with R_t, which thus appears as the antenna's source resistance. If the $\lambda/2$ antenna is attached to a matched load, then in accordance with the maximum power theorem, half the antenna's open circuit terminal emf will appear across the load and as much energy is dissipated internally in the source as in the load. Unlike a conventional signal source, however, the power dissipated in the antenna does not appear as heat (assuming R_l is small), but is reradiated by the antenna as a spherically expanding wave with both near- and far-field components. Thus in the immediate vicinity of the antenna, the resultant field is due to the combination of the original plane wave and the spherical reradiated wave.

The maximum amount of energy which a loss-free receiving antenna can deliver to a matched load is related to its 'effective aperture' A, an area at right angles to the direction of propagation of the signal. A lossless isotropic antenna has an effective aperture $A = \lambda^2/4\pi$, thus A is a function of the wavelength and does not depend upon the physical size of the antenna. For practical antennas, $A = G\lambda^2/4\pi$, where G is the power gain of the antenna; thus a lossless dipole has an effective aperture $A = 1.65\lambda^2/4\pi$.

In many situations, from a VHF or UHF pocket pager to a military tactical HF communications system, size or weight considerations may enforce the use of an antenna that is much smaller than a half-wave dipole. Such an antenna will not be resonant in its own right, but measures can be taken to bring it to resonance. For example, a $\lambda/4$ dipole can be fitted with end discs, like the ends of a soft drinks can. Where the size is even smaller relative to a wavelength, either a loop or a dipole can be used and tuning components built in to bring it to resonance (Figure 12.9). However, with an electrically very small antenna, the radiation resistance becomes very low, with two important consequences. Firstly, as the ratio of the antenna's reactance to R_l is high, when brought to resonance the Q will be high, giving a very narrow useable percentage bandwidth. Secondly, R_l

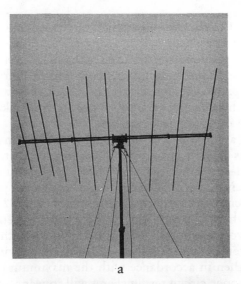

a

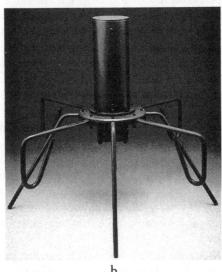

b

Figure 12.7   *Antennas.*
a VHF log periodic directional antenna, rated 30–88 MHz, 400 W. For lightness, economy and ease of transportation, the longer elements are loaded, allowing their physical length to be less than their electrical length.
b RA978 UHF ground-to-air omnidirectional monopole antenna, rated 220–400 MHz, 1.2 kW pep. Available in both CAA and NATO codified versions.
(Reproduced by courtesy of Racal Antennas Ltd.)

a

Figure 12.8

a Skanti TRP8750 750-W HF SSB radio telephone transreceiver with (at left) its ATU. This is designed to drive an unbalanced antenna (long wire, inverted L, whip, etc.). It is fully auto-tuning under control of its internal microprocessor.

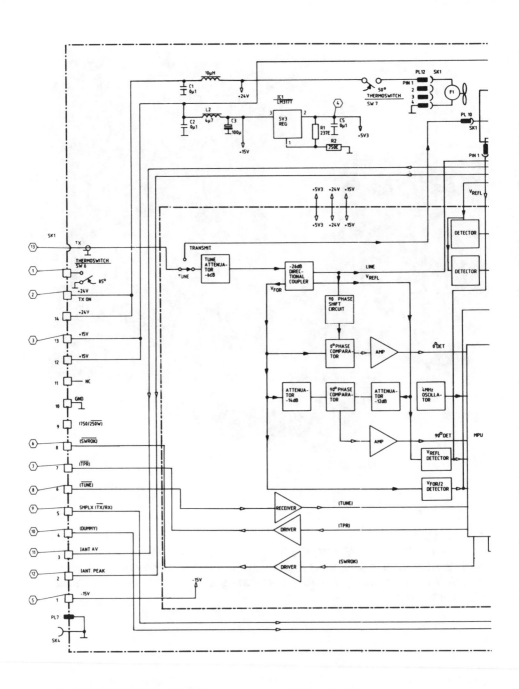

Figure 12.8   Continued
*b Block diagram of the antenna tuning unit of TRP8750 T/R.
(Reproduced by courtesy of Skandinavisk Teleindustri Skanti A/S.)*

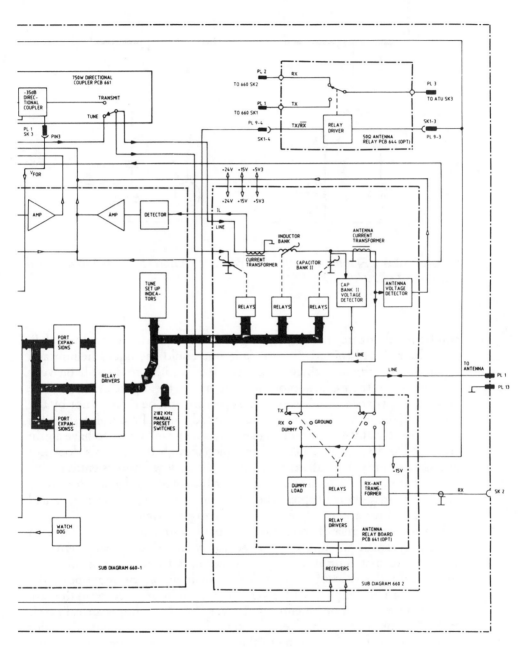

b

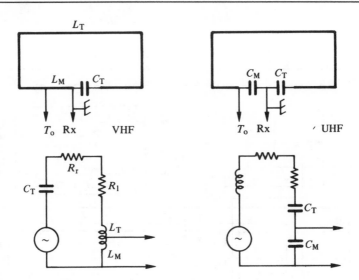

Figure 12.9 *Electrically small antennas, tuned and matched, with equivalent circuits.*

will be much greater than R_r leading to a very low efficiency. Even if R_l could be reduced to zero (in principle one could use liquid helium and superconductivity to achieve this), the bandwidth would still be very narrow due to the high ratio of the reactance of the dipole or loop to the radiation resistance R_r. However, the aperture will be defined not by the physical size but by the wavelength, as noted above. Practical designs for passive electrically-small receiving antennas may well prove to have a gain G up to 20 dB or more below isotropic (though though this does not necessarily apply to small *active* antennas). This low figure is entirely due to the loss resistance R_l, a small dipole or loop will still have a *directivity* or gain-relative-to-isotropic. The literature covering electrically small antennas, which are mainly used for receiving, is extensive [5, 6].

The foregoing relates to electrically small *passive* antennas. Where an electrically small antenna is intended for receiving only, an alternative approach to matching it directly to a feeder, is to design it as an active antenna. In the case of an electrically small dipole or monopole, the amplifier can be designed with a very high input impedance, or in the case of a small untuned loop antenna, with a very low input impedance, in each case the amplifier output being designed to match a standard feeder impedance, such as 50 Ω. Due to the small aperture of such an antenna, and the lack of matching, the signal energy available to the amplifier will be small, but provided it exceeds the amplifier's internal noise by a sufficient margin, this will still allow satisfactory operation. In consequence, active antennas are particularly useful in the LF, MF and HF bands, where external noise greatly exceeds thermal noise, and is thus well above the internal noise of a suitably designed amplifier. Active antennas are offered by a number of manufacturers, in many cases the internal circuit design being a proprietary secret. Figure 12.10 shows an active HF antenna which, though no longer in the catalogue, is typical of the design of these antennas.

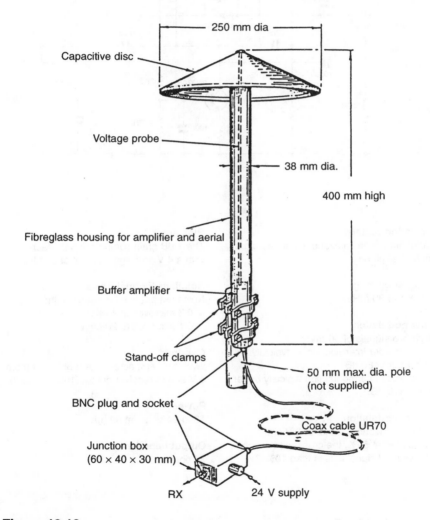

Figure 12.10
a An active HF antenna, showing its general mechanical arrangement, and its power-insertion junction box.

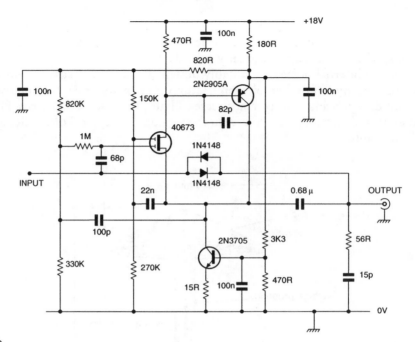

b

Radiation pattern
Omni-directional in azimuth, semi-toroidal in vertical plane.

Frequency range
10 kHz to 30 MHz.

Intermodulation
With two signals of 30 mV:
Second order intermodulation typically better than −80 dB
Third order intermodulation typically better than −110 dB.

Cross modulation
With an unwanted signal of 2 V emf, modulated at 50%, the cross-modulation of a wanted signal is less than 10%.

Blocking
The 1 dB grain compression is reached with a 4 V emf signal output at 30 MHz.

Amplifier thermal noise
Noise out put in 6 kHz bandwidth:
0·3 microvolt at 1 MHz
0·1 microvolt at 20 MHz.

Overload
With 30 V emf across the probe maximum 5 V emf to receiver output (100 V/m field).

Power
18 to 24 V, dc, at 50 mA

Output impedance
75 ohms.

c

Figure 12.10 Continued
  b *Circuit diagram of the antenna.*
  c *Summary of performance characteristics.*

An active antenna such as that just described is effectively operated by the E field component of the signal. If an electrically small antenna must be situated in a position where it is subject to electrostatic interference, a loop antenna — which is operated principally by the H field of the signal — may prove more suitable. Figure 12.11, reproduced from Reference 7, shows such an active loop antenna, with gain switchable between about 8 dB or 20 dB. A three turn 15 inch diameter coil of 8AWG wire with $\frac{1}{2}$ inch turns spacing tuned with a dual gang 10–330pF capacitor covers 4.4 to 16 MHz. A single turn coil made by bending a 48 inch long strip of $1\frac{1}{4}$ inch wide Ali sheet into a circle will cover from 13 MHz to beyond the top of the HF band, being useful at reduced performance right up to 55 MHz.

Commercial loop antennas are available, offering very high rejections of electrostatic interference. These use a loop where the turn(s) are enclosed in an earthed screening tube. A short gap in the tube prevents its presenting a shorted turn, enabling the H field to induce an emf in the inner, whilst screening the antenna from any electrostatic interference.

Transmitting antennas are usually required to have a higher efficiency than that which may be acceptable in a receiving system. Nevertheless, the laws of physics are immutable and one may have to accept an efficiency as low as a few per cent in the case of a tactical HF antenna at the lower end of the band. Such an antenna is 'broadbanded' by including load resistors

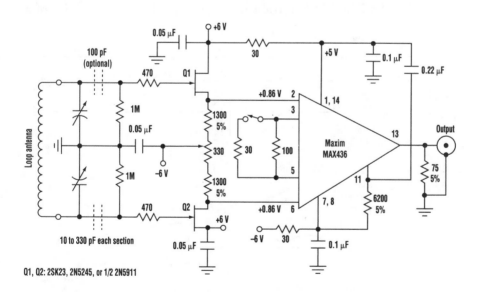

Figure 12.11 *A high-frequency loop antenna. Reprinted with permission from Electronic Design, July 22, 1996, Copyright 1966, Penton Publishing Co.*

which play no part at the higher frequencies where the antenna is not electrically small, but which keep the transmitter happy by maintaining the antenna's VSWR within limits (e.g. less than 2.5:1) in the 2–4 MHz region where it is small in relation to λ. One such well-publicized antenna, popular with amateur radio operators, is shown in Figure 12.12: it is commonly known as the 'Australian dipole' and has also been tested and used by government agencies and commercial firms. With its overall length of 40.4 m, it is in fact only about 20% shorter than a halfwave dipole at its lowest rated frequency of 3 MHz, its main advantage being that it maintains a VSWR of 2.5:1 or better from there up to 30 MHz. But whilst presenting a reasonable match to a transmitter at all operating frequencies, ensuring that much of the available power is radiated, its actual radiation pattern is another question entirely. In azimuth it will be figure-of-eight, while the elevation pattern will depend upon the height at which it is mounted, and the frequency of operation. But in general, the elevation pattern will be multi-lobed at higher frequencies. It will be clear from the earlier discussion of antenna mounting height, that the actual antenna gain or loss relative to isotropic at any given elevation angle, at any frequency, will be somewhat uncertain, even varying with the degree of wetness of the ground in the vicinity.

Another electrically small transmitting antenna which has created some interest in recent years is the 'crossed field antenna'. It has been noted earlier that the E and H (electric and magnetic) fields in the vicinity of a

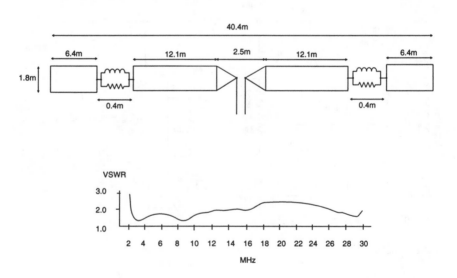

Figure 12.12 *The 'Australian dipole' exhibits a VSWR of no worse then 2.5:1 over its operating range of 3 to 30 MHz.*

dipole are in quadrature phase, so representing stored energy in what is effectively a tuned circuit, whereas radiation is only evidenced by the far field, where the E and H components are in phase, mutually orthogonal, and both orthogonal to the direction of propagation, as described by the Poynting vector. The crossed field antenna aims to synthesize the Poynting vector by producing separately stimulated E and H fields, and superposing them in the same 'inter-action space' around the antenna, to produce a radiated power flux S = ExH, where the x indicates a vector cross-product. The input power is split, and half applied to a pair of electrodes designed to produce the required E field pattern. The other half is used to produce a corresponding H field. One version of the system which been described in the literature is said to cover 1.8–28 MHz, although it should be stressed that this is not the instantaneous bandwidth. The latter typically varies from about 100 kHz at 3.65 MHz to 400 kHz at 21 MHz, the elements of the splitter and phasing units requiring readjustment when the operating frequency is changed. The performance of the system is claimed to be good, but in view of its unorthodox approach this is disputed by most proponents of more conventional antennas.

So far, only simple antennas, dipoles, monopoles, loops, etc., have been considered. Antennas with several elements can provide greater directivity than a dipole and thus exhibit an aperture (as far as transmission or reception in the preferred direction is concerned) of greater than $1.65\lambda^2/4\pi$. Antenna power gains G of up to 10 or 20 times (10–13 dB) are possible in HF antennas. Such high gain antennas are usually restricted to a fixed direction of operation, due to their size, but rotatable high gain HF antennas are available . (One type of antenna suitable for this purpose is the 'log-periodic' antenna, see Figure 12.7a, a multi-element antenna which can be designed to cover a relatively wide bandwidth.) This naturally presupposes one knows where the other end of the link is: for a more fluid situation, e.g. ground/air communications, or where messages must be broadcast to several vehicles, both ends of the link are likely to employ antennas designed to be as nearly omnidirectional as possible — no easy task on an aircraft. At VHF, gains in excess of 20 dB are possible, using array antennas such as stacked Yagis. (The Yagi antenna, which is narrow band, consists of a half-wave dipole plus parasistic elements which modify the pattern; a reflector behind the main element and a number of directors in front of it.)

For a thin wire half-wave dipole, the aperture of $1.65\lambda^2/4\pi$ square metres seems to bear little resemblance to the actual area, which is clearly much less than this. However, with a large antenna array, or a dish antenna, where the overall dimensions may be many wavelengths, it is found that the actual physical area does approximate to the effective area $A = G\lambda^2/4\pi$. For example, a microwave dish of physical area a, will have an effective area of $A = 0.6a$, approximately. (The factor 0.6 is due to the impossibility of designing a feed system which will distribute the power uniformly over the reflector without spilling any over the edge.) Thus at microwave where dish antennas are commonly employed, gains of 40–50 dB are available.

In all cases of directional antennas, the increased gain in the desired direction is bought at the expense of reduced gain in other directions. With high gain antennas, there are usually a number of 'sidelobes', directions in which the gain, though much less than that in the main lobe, is nevertheless considerable. In some cases a directional antenna is employed more to discriminate against unwanted signals coming from a different direction from the wanted signal, than to increase the gain to the latter. A common example is in TV reception, where an antenna with a high front-to-back ratio can reduce ghosting due to reflections of the wanted signal, or interference due to another station. The examples just given are mostly terrestrial situations; only in space applications or in microwave links using very directional dish antennas will the free-space path loss formula be applicable.

So far, only individual antennas have been considered. The chapter would not be complete, however, without some mention of antenna arrays. These may be used for a number of purposes. For instance, an in-line array of antennas, all fed with equal amounts of power in the same phase from a transmitter via a splitter, will produce narrow beams, like a long thin figure-of-eight at right angles to the array, plus various sidelobes. On the other hand, if each individual antenna is fed with the signal, in equal amounts but suitably successively delayed in phase, a narrow end-fire beam is effected. Such linear arrays, given the necessary adjustable phasing arrangements, can be used as directional receiving antenna systems, and hence also as DF (direction finding) systems. Circular arrays of monopoles are used in DF systems at HF, such as in the Wullenweber system (where the large aperture permits the synthesis of narrow beams, especially in the upper part of the HF band) and at VHF, e.g. short range coastal DF installations. Compact arrays are necessary where space is limited, e.g. the Bellini-Tosi antenna (consisting of crossed triangular loops connected to a goniometer) once commonly used for ship-borne DF. Another example is the Adcock DF antenna (consisting of four vertical half-wave dipoles mounted at the ends elevated cross-arms and connected to phase-difference measuring equipment) for tactical DF applications, where rapid re-deployment is a requirement.

A major accuracy limitation in DF systems, at both HF and VHF, is due to the reception of different rays, i.e. different versions of the same signal via different paths. At HF, these will usually be different skywave paths, whilst at VHF there may be both direct and reflected rays; both are examples of multipath propagation. In addition, in the tactical environment there is often great interest in DF on co-channel signals. Modern techniques, employing multichannel receivers fed from an antenna array, and feeding into DSP (digital signal processing) equipment can provide enhanced performance in these circumstances. Various algorithms are used, all under the general heading of 'SuperResolution DF', giving the ability to resolve two or more simultaneous co-channel signals whose angular separation is less than the natural beamwidth of the array. SR-DF does not rely on any particular array geometry; it is only necessary to know the relative positions of the antennas and their patterns. The technique requires only a few samples of the signal to provide an accurate bearing, and is thus suitable for use in 'ambush mode' (monitoring a particular frequency in the band) for DF on frequency hopping transmitters.

Finally, there are specialized antennas for field-strength measurements. These are covered in Chapter 14, Measurements.

References

1. Kraus. *Antennas*, McGraw-Hill, New York (1950)
2. Jasic (ed.) *Antenna Engineering Handbook*, McGraw-Hill, New York (1961)
3. Schelkunoff and Friis. *Antennas, Theory and Practice*, John Wiley and Sons, New York (1952)
4. Terman. *Radio Engineering*, 3rd edn, McGraw-Hill, New York, p. 716
5. Virani. Electrically small antennas. *Journal I.E.R.E.*, **538**(6), 266–74 (Sept–Dec 1988)
6. Fujimoto *et al. Small Antennas*, Research Studies Press
7. Salvati. High-Frequency Loop Antenna, *Electronic Design*, July 22, 1996

13
Attenuators and equalizers

Attenuators, or pads as they are often called, are networks which simulate a lossy transmission line, so that the signal at the output is smaller than at the input, but not changed in any other way. Like a transmission line, they are designed to have a specific characteristic impedance, commonly 50 Ω, and like a good transmission line their frequency characteristic is flat. Unlike a length of lossy line though, they provide no delay; the path length through an attenuator is ideally zero. A pad exhibiting a resistive impedance R_0 at both its input and its output can be realized with three resistors connected in either a 'Tee' or a π configuration (see Figure 13.1), which gives design formulae expressed in two different ways. The first gives the hyperbolic design equations for the series and shunt resistors of a Tee pad in terms of the attenuation α in nepers where $\alpha = l_n E_{in}/E_{out}$, i.e. the natural logarithm of the voltage ratio. The second way uses the input/output voltage ratio N where the required attenuation D dB is given by $D = 20 \log_{10} N$. You can thus work out the resistor values for a pad of any attenuation for any characteristic impedance, but for most attenuation values for common characteristic impedances such as 50 Ω or 600 Ω it is quicker to look up the values in published tables, such as Appendix 3. Note that if the voltage (or current) ratio is very large, then (1) the coupling between input and output circuit must be very small, and (2) looking into the pad from either side we must see a resistance very close to R_0 even if the other side of the pad is unterminated. For if very little power crawls out of the far side of the pad, it must mostly be dissipated on this side. Thus when N is very large, (1) R_p in a Tee circuit must be almost zero and R_s in a π circuit almost infinity, and (2) R_s in a Tee circuit will be fractionally less than R_0 and R_p in a π circuit fractionally larger than R_0. In fact as you can see from Figure 13.1b, the R_s in a Tee circuit is the reciprocal of R_p in a π circuit (in the sense that $R_{s(Tee)} R_{p(Pi)} = R_0^2$) and vice versa, for all values of N. Figure 13.1c shows eight switchable pads arranged to give attenuation in the range 0–60 dB in 1 dB steps. The range can be extended by adding further 20 dB sections,

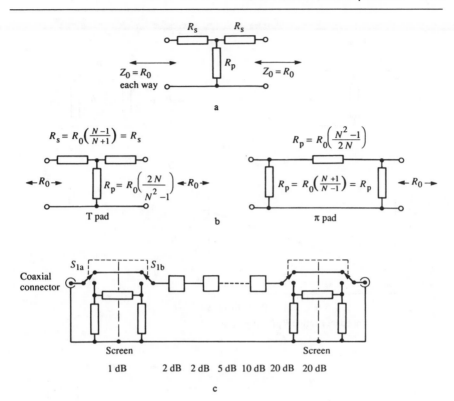

Figure 13.1 *Attenuators.*
a Attenuator design in exponential form: R_s = $R_0 \tanh \alpha/2$, R_p = $R_0/\sin h\alpha$, true for all α (in nepers).
b Attenuator design in terms of input/output voltage ratio N: attenuation D = 20 $\log_{10}$ N dB.
c 0–60 dB attenuator with 1 dB steps.

or by adding a 40 dB section. However, in practice the former permits operation up to much higher frequencies, since with attenuations in excess of 20 dB in a single pad, worse errors due to stray capacitance and inductance will be encountered.

A variable attenuator is useful for many measurement applications. Continuously variable attenuators using resistive elements have been designed and produced but are expensive, since three resistors have to be varied simultaneously, with non-linear laws. Continuously variable attenuators working on a rather different principle are readily available at microwave frequencies. Piston attenuators, working on the waveguide beyond cut-off principle are also available for use at V/UHF. Alternatively, attenuators

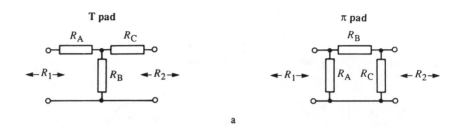

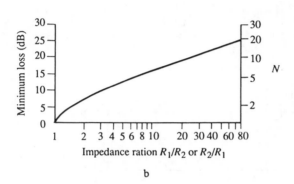

a

b

Figure 13.2
  a *Mismatch pads.*

$$R_B = 2\,R_0 \frac{N}{N^2 - 1}$$

$$R_A = R_1 \frac{N^2 + 1}{N^2 - 1} - 2\,R_0 \frac{N}{N^2 - 1}$$

$$R_C = R_2 \frac{N^2 + 1}{N^2 - 1} - 2\,R_0 \frac{N}{N^2 - 1}$$

T pad
$R_0 = \sqrt{(R_1 R_1)}$

$$R_B = \frac{R_0}{2} \frac{N^2 - 1}{N}$$

$$R_A = R_1 \frac{N^2 - 1}{N^2 - 2NS + 1}$$

$$R_C = R_2 \frac{N^2 - 1}{N^2 - 2(N/S) + 1}$$

π pad
$R_0 = \sqrt{(R_1 R_2)}$
$S = \sqrt{(R_1 / R_2)}$

  b *Minloss pads.*

adjustable in 1 dB steps are modestly priced and very useful. For example, if the output of a signal generator is measured with an indicating receiver of some sort, and then an amplifier in series with the attenuator is inserted in the signal path, then when the attenuator is set to provide the same receiver indication as previously, the amplifier's gain equals the attenuator's attenuation. The accuracy of the measurement depends only upon that of the variable attenuator, not on the source or detector. The output of the signal generator should not of course be large enough to drive the amplifier into saturation: if, due to limited detector sensitivity, it is necessary to work with a signal level larger than the amplifier can handle, the attenuator can precede rather than follow the amplifier.

Fixed pads are useful for providing some isolation between stages, albeit at the expense of a power loss. In particular, the use of a pad will reduce the return loss of a poorly matched load seen by a source, or vice versa (see Appendix 3). Sometimes it desired to connect together two systems with different characteristic impedances, to measure the performance of a 75 Ω video amplifier using a 50 Ω network analyser, for example. Impedance matching transformers could be used for this purpose, but their frequency range might prove inadequate. A much broader-band solution is to use a pair of 'mismatch pads' (a palpable misnomer — they are actually 'anti-mismatch pads'). A 50 Ω to 75 Ω pad would be used at the amplifier's input and a similar pad, the other way round, at its output. Figure 13.2 gives the design formulae for both T and π mismatch pads; note that here N is not the input/output voltage ratio but the square root of the input/output power ratio. For any ratio of impedances to be matched there is a minimum associated loss, e.g. for a pair of 1.5 : 1 pads (75 Ω to 50 Ω for example), from Figure 13.2b the loss cannot be less than about 6 dB, unless that is you resort to the use of negative values of resistance in which case you can have a 0 dB mismatch pad or even one with gain. In practice, it is convenient to design the pads for say 10 dB each so that the actual gain of the 75 Ω video amplifier mentioned above would be 20 dB greater than the measured value. If the above set-up were being used to measure the stopband attenuation of a 75 Ω filter, the extra 20 dB loss of the mismatch pads would undesirably limit the measurement range. In this case it would be better to use 'minloss' pads. These are L pads, having only two resistors, a series resistor facing the higher impedance interface and a shunt resistor facing the lower.

Whereas an attenuator provides a loss that is independent of frequency and a filter has an attenuation that varies with frequency, a phase equalizer has no attenuation at any frequency. For this reason it is alternatively known as an all-pass filter (APF) and it is used to provide a phase shift that is dependent upon frequency. A typical application is in a digital phase modulation system where an LC or (more usually) active RC low-pass filter is used at baseband prior to the modulation stage, to limit the bandwidth of transmitted signal. An APF can be used to correct the phase distortion introduced by the baseband filter. The aim is to make the phase shift through the filter/equalizer combination linearly proportional to frequency: when this 'constant group delay' condition is met, all frequency components of the digital data stream suffer the same time delay and so their relative phase is unaffected, avoiding ISI (intersymbol interference) in the transmitted signal. The overall link filtering function is usually split equally between the transmitter and the receiver, to obtain the best trade-off between OBW (occupied bandwidth) of the transmitted signal and noise bandwidth at the receiver. However, all of the corresponding equalization may be carried out at one end of the link, say the transmitter, if convenient. A first order phase equalizer provides a phase shift which increases from zero at 0 Hz to 180° at frequencies much higher than its designed 90° centre frequency, the phase variation versus frequency being of a fixed shape. A second order section provides a phase shift which increases from zero at 0 Hz to 360° at frequencies much higher than its designed 180° centre frequency; the rapidity of phase change in the region of the centre frequency being a variable at the disposal of the designer. An equalizer having a number of sections will usually be necessary to equalize the baseband filter. Both first and second order APF sections are described in Reference 1. Phase equalization is not necessary if the baseband filter has a constant group delay, i.e. phase shift proportional to frequency throughout the pass-band. Among *LC* filters, the best known design possessing this property is the Bessel filter, but its rate of cut-off is too gradual to provide the desired degree of bandwidth limitation. Linear phase filters with a sharp cut-off at the band edge can be realized using capacitors and inductors [2] by adopting a non-minimum phase design. Reference 3 describes how a low-pass version of such a filter can be realized using an active RC approach. Finite impulse response (FIR) filters exhibit an inherently linear phase/frequency characteristic and they are available either in DSP (digital signal processing) implementations, or as charge-coupled devices.

It was mentioned in an earlier chapter that a double balanced mixer used as the first mixer in a high grade receiver should ideally see a broadband 50 Ω termination at each of its three ports. Often it is not possible to arrange for this desirable state of affairs, but it can be approached. The local oscillator port can be driven by an amplifier with a broadband resistive output and it may prove possible to drive the RF port from a low-gain buffer amplifier to isolate it from the large out-of-band VSWR of the RF band-pass filter. A broadband match at the IF port is more difficult to achieve but it can be approximated by a frequency selective constant resistance network. Such networks have many uses, a familiar domestic example being the cross-over network used to direct the low frequency and high frequency parts of the output of a hi-fi system to the woofer or the tweeter respectively. Figure 13.3 shows a constant resistance band-pass filter network which preserves a constant 50 Ω resistive characteristic at both input and output port in its stop

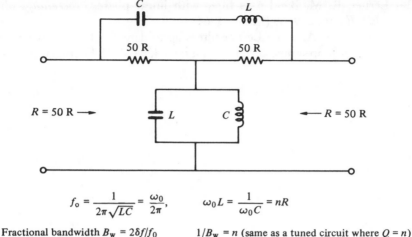

$$f_0 = \frac{1}{2\pi\sqrt{LC}} = \frac{\omega_0}{2\pi}, \qquad \omega_0 L = \frac{1}{\omega_0 C} = nR$$

Fractional bandwidth $B_w = 2\delta f/f_0$ $1/B_w = n$ (same as a tuned circuit where $Q = n$)
-3 dB at $f_0 \pm \delta f$

Figure 13.3 *Constant resistance band-pass filters.*

bands. The pass band is centred on frequency $f = \{2\pi\sqrt{(LC)}\}^{-1}$ and the higher the L/C ratio, the narrower the pass band. However, the higher the value of inductance used, the higher the required Q if the pass band loss is to be kept low. Assuming the pass-band loss is low, the network is transparent in its pass band, so that the VSWR at its input is simply that of the load on the network's output. If this is an IF crystal roofing filter, the input VSWR of the network plus roofing filter will be low in the latter's pass band, but will rise at greater frequency offsets, until it finally falls again in the stop band of the constant resistance network. The poor VSWR immediately either side of the crystal filter's pass band is unfortunate, but the arrangement is still a considerable improvement upon a direct connection of the crystal filter to the mixer. Alternatively, a high reverse isolation buffer amplifier with low return loss at both input and output ports may be interposed between the constant resistance network and the crystal roofing filter. The latter now sees a good match at all frequencies, both in and out of band. The constant resistance band-pass filter protects the buffer amplifier from the welter of out-of-band signals at the mixer's output port, while the latter is now correctly terminated at all frequencies.

References

1. Hickman, I. *Analog Electronics*, Heinemann Newnes, Oxford, pp. 128–50 (1990)

2. Lerner, R. M. Band-pass filter with linear phase. *Proceedings of the I.E.E.E.*, pp. 249–68 (March 1964)
3. Delagrange, A. Bring Lerner filters up to date: Replace passive components with op-amps. *Electronic Design*, **4**, 94–8 (15 February 1979)

14
Measurements

In any serious development work, evaluation or production test in connection with RF equipment, suitable test equipment is a must, a *sine qua non*. With it, one can measure the frequency, amplitude and phase noise of a CW signal and the relative levels of any harmonics present, the AM, FM or PM modulation on a signal modulated by a single sinewave, or the characteristics of more complex types of modulation such as the various forms of phase shift keying, stereo FM or television signals, etc. Without it, one is working in the dark. This chapter looks at the types of equipment needed to make measurements on the above signals, and also at making measurements on circuit parameters, such as the frequency response, input and output VSWR of amplifiers, and the s-parameters of RF amplifiers, etc. Then there is also the question of the measurement of signals in space, i.e. field strength measurements. These are required not only for determining whether a particular comunications link is viable — for example where to place a TV antenna to obtain an adequate picture free of ghosting or interference from other stations — but also checking that the out-of-band emissions from a transmitter are within the limits permitted by current legislation.

Measurements on CW signals

The amplitude of a CW signal may be measured in many ways, one traditional instrument being an RF millivoltmeter. These used a diode detector and could measure signals in the range (typically) 10 kHz to 1 GHz. They typically had a high input impedance and so could be tapped across an RF line to make a 'through' or 'bridging' measurement with minimal disturbance to the circuit under test, or used in conjunction with a 50 Ω termination for terminated measurements. The measured value with such an instrument could be affected by the presence of odd order harmonics and, in many cases, even order harmonics also, so their

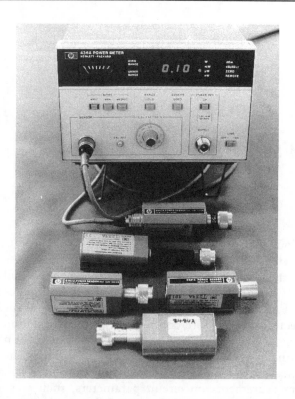

Figure 14.1 *The Hewlett-Packard model 436A power meter with a variety of sensors. With a frequency range of 100 kHz–50 GHz and a power range of –70 to +44 dBm (both sensor dependent), the HP 436A measures both absolute and relative power.*
(Reproduced by courtesy of Hewlett-Packard Co.)

popularity has waned. Current RF voltmeters are mainly of the true rms variety, though high input impedance types tend to be limited to the HF range and below, e.g. Marconi Instruments 2610 (25 MHz) or Fluke/Philips 8290A (20 MHz). For higher frequencies, terminating (50 Ω or 75 Ω) true rms power meters are normally used. The sensors may be thermocouples, or diodes operated at a very low level — where their response is rms rather than linear. Typical examples are the Marconi Instruments 6960A, and the Hewlett-Packard HP436A which is illustrated in Figure 14.1.

The determination of the exact frequency of an RF signal was in former days a complicated business but is nowadays simply a matter of connecting it to a digital frequency meter. A frequency counter function is built in to many general purpose DMMs (digital multimeters), such as the

Philips PM2525 (10 Hz–20 MHz), whilst bench-top timer/counter/ frequency meters offer a wider range. A typical example is the Philips PM6665 which measures frequencies up to 1.3 GHz via a 50 Ω terminated input and up to 120 MHz via a 1 MΩ/35 pF high impedance input.

The phase noise of a CW signal can be measured in various ways, the simplest being to use a high grade spectrum analyser. The harmonics of an RF signal can also be measured with a spectrum analyser. This is such a versatile instrument that it is covered in detail later in the chapter.

Modulation measurements

For the measurement of AM, FM or PM the most convenient instrument is a modulation meter. In addition to measuring the modulation depth or deviation, most modulation meters will also make a high-quality demodulated output available for monitoring purposes, and additionally make measurements such as carrier frequency and level, frequency response, signal to noise ratio, stereo separation, etc. It is possible to measure the AM of a signal which also carries FM (or PM) and vice versa. Usually, in addition to manual tuning, an auto-tune function is available to instantly tune the instrument to the only (or largest) carrier present. Typical examples are the Marconi Instruments 2305 (50 kHz–2.3 GHz carrier range) and the Hewlett-Packard HP8901A/B (150 kHz–1300 MHz). Both of these instruments, like most others mentioned in this chapter, are capable of being controlled over the GPIB instrument bus.

Special instruments are available for measuring the parameters of signals bearing the more complicated sorts of modulation. A good example is the Hewlett-Packard HP 8980A, which accepts the I and Q channel outputs from an I/Q demodulator and displays the constellation diagrams of high-rate modulation schemes such as QPSK, 16QAM, 49PRS, etc. It also makes statistical measures of system quality like eye closure, lock angle error and quadrature error (see Figure 14.2). This permits evaluation of a receiver/demodulator system when driven from a suitable source, such as an HP 8780A Vector Signal Generator. The HP 8981A is similar to the HP 8980A but with the addition of a built-in I/Q demodulator; it can thus accept and test the IF I/Q output of a modulator.

Another interesting and useful approach to complex modulation measurements is offered by the Hewlett-Packard HP 5371A (accepts input frequencies up to 500 MHz) and HP 5372A (up to 2 GHz) Modulation Domain Analysers. The input frequency range can be extended to 18 GHz using the HP 5364A Microwave Mixer/ Downconverter. Whereas a time domain instrument such as an oscillo-scope displays amplitude (instantaneous voltage) versus time and a frequency domain instrument — the spectrum analyser — displays rms amplitude versus frequency, the modulation domain analyser can display

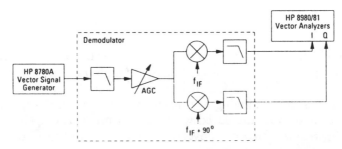

The HP 8980A Vector Analyzer and the HP 8780A Vector Signal Generator can be used to adjust and troubleshoot an I/Q demodulator directly. The I/Q outputs of the demodulator are connected directly to the HP 8980A. The HP 8980A Vector Analyzer can display the constellations of high-rate modulation schemes such as QPSK, 16QAM, 49PRS, 64QAM, and 56QAM. It also makes statistical measures of system quality like closure, lock angle error and quadrature error.

a

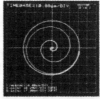

UMOP (Unintentional-Modulation-on Pulse) is identified by quantitatively measuring the phase transients on a radar pulse with the delta-phase measurement marker.

Display of vector demodulated SAW chirp signals. The spiraling phase response indicates the changing chirp frequency and amplitude.

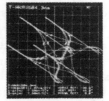

I & Q display: each I and Q channel is displayed vs. time on a separate grid, one above the other.

Constellation display: displays Q vs. I at the time instant defined by the time marker.

3D display: useful for visual, or intuitive, analysis of Q vs. I vs. time waveforms. Signal can be rotated about any of three axes for optimal viewing.

b

Figure 14.2   *The Hewlett-Packard vector analysers HP8980 and HP8981 can make a variety of measurements on complex signals. (Reproduced by courtesy of Hewlett-Packard Co.)*

frequency versus time, phase versus time, or time interval versus time. This makes it simple to display and analyse otherwise tricky measurements such as chirp linearity in a radar system, settling time of a VCO output following a step frequency change in a PLL synthesizer, or the frequency hopping performance of a frequency-agile transmitter. Frequency versus time can be measured by examining the detected output of a spectrum analyser used in zero-sweep mode, but the measurement capability is severely limited in bandwidth by the spectrum analyser's IF circuits. Being basically a counter-based instrument, this restriction does not apply to the Modulation Domain Analyser, which can cope with signals no matter how wideband they are.

Spectrum and network analysers

These instruments are so fundamental to the RF engineer that they deserve a section to themselves. The spectrum analyser is a development of the earlier panoramic receiver, which was a swept receiver displaying the amplitude of any signals it encountered within the frequency range over which it was swept. Apart from greater stability and selectivity, the main difference is that the modern spectrum analyser can display the signals on a logarithmic scale covering (typically) 80 dB at 10 dB per vertical division. Additionally, for finer amplitude discrimination, a vertical scale of 2 dB/ division and also a linear scale are usually available. Manufacturers of spectrum analysers include Hewlett-Packard, Tektronix, Marconi Instruments, Advantest (Takeda Riken), Anritsu, Rohde & Schwarz, Wandel & Goltermann and a number of others. Models from Tektronix cover frequencies up to 33 GHz, and models in the 2380 series from Marconi Instruments cover frequencies up to 26.5 GHz. Figure 14.3 shows the model 2382 from Marconi Instruments: the attractively economical price of this instrument is achieved by limiting the range to 400 MHz. Combined with a 100 dB on screen display range (at 10 dB/division), it offers resolution bandwidths from 1 MHz down to 3 Hz in a 1, 3, 10 sequence and a video (post-detector) bandwidth of 50 kHz down to 1 Hz may be selected. (This instrument was used to produce Figures 7.1e, 7.2 and 10.6.)

A spectrum analyser may be used for a wide range of measurements, including determining the relative amplitude of any harmonics of an RF signal. It may also be used to measure the phase noise (sideband noise) of an unmodulated carrier, provided of course that the phase noise of the spectrum analyser itself is lower than that of the CW source under test. Another important test conveniently carried out using a spectrum analyser is intermodulation testing. A typical application is testing the linearity of an HF SSB transmitter, by the two-tone test method. Here, two equal amplitude audio-frequency tones, say 1000 Hz and 1700 Hz,

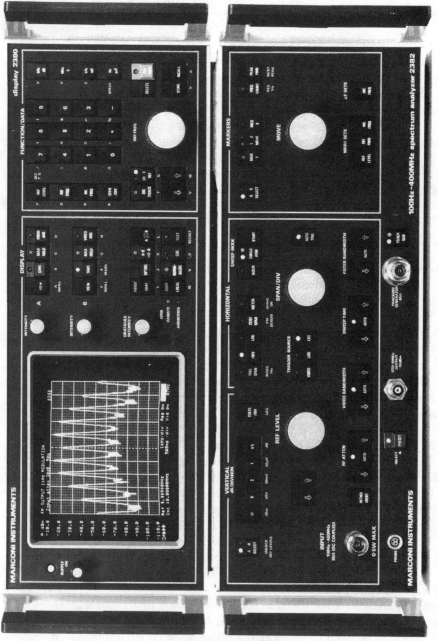

Figure 14.3 The Marconi instruments spectrum analyser 2382 covers the frequency range 100 Hz to 400 MHz. Other models in the 238X range cover higher frequencies – up to 26.5 GHz. (Reproduced by courtesy of Marconi Instruments Ltd.)

are combined and applied to the transmitter's modulation input, taking care to isolate each tone from the other so that intermodulation does not occur between them, e.g. in the tone generators' output circuits. A sample of the transmitter's output is then applied to the spectrum analyser, and if no intermodulation has occurred, the only signals found will be (assuming for example USB modulation) two equal amplitude components at 1000 Hz and 1700 Hz above the suppressed carrier. In practice, the carrier suppression will not be complete, though the usual specification calls for it to be at least 40 dB down on PEP (peak envelope power).

In the two-tone test, assuming that intermodulation is not severe, PEP will be 6 dB above the level of either of the two RF tones. If third order intermodulation occurs in the transmitter, as is bound to be the case to some extent, additional components will be seen in the output, offset by the separation between the tones, e.g. at 700 Hz above the higher frequency tone and at 700 Hz below the lower. The permitted level of these tones depends upon the applicable specification, as published by the FCC (Federal Communications Commission, applicable in the USA), ITU-R (International Telecomnications Union, Radiocommunication Bureau, formerly known as CCIR — International Radio Consultative Committee), or whatever.

The relevant ITU-R specification is Recommendation 326, and this has been embodied in the national regulations of many European companies. This specificaton calls for the third order intermodulation products in an HF SSB transmitter operating in J3E mode (formerly known as A3J mode) in normal speech service to be 26 dB down on either of the two tones. The earlier versions of Recommendation 326 were unfortunately worded in such a way that the requirement could be interpreted as being 26 dB down on PEP. My suggested re-wording was submitted to the ITU by CCIR U.K. Study Group 1, ratified by a Plenary Assembly, and is incorporated in the current version. The requirement for transmitters where a privacy device is fitted is tighter, at 35 dB down on either tone. The higher figure is because a device such as a scrambler will disperse the speech energy throughout the sideband, resulting in a greater likelihood of significant intermodulation products falling into adjacent channels. Both carrier suppression and IMP (intermodulation products) are quickly and simply tested with a spectrum analyser.

Another instrument important to the RF engineer is the network analyser. This measures the analogue characteristics of electronic products including components, circuits and transmission lines. Consequently it is widely used in many fields from R&D to mass production, for analysing the transmission, reflection and impedance characteristics of these products. Manufacturers of network analysers are much fewer in number than those of spectrum analysers. Further, some manufacturers of network analysers produce only scalar instruments, rather than the more

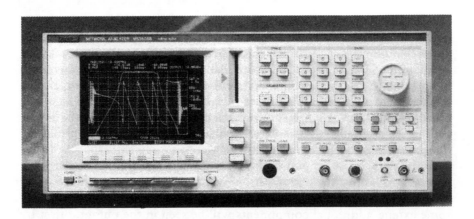

Figure 14.4 *The Anritsu MS3606B network analyser makes scalar, vector (magnitude and phase) and group delay measurements over the range 10 kHz–1 GHz. The screen display shows the magnitude and phase response of a 10.7 MHz filter. The constant slope of the phase characteristic in the passband indicates a constant delay characteristic. See also Figure 6.5.*
(Reproduced by courtesy of Anritsu Europe Ltd.)

generally useful vector instrument. Basically, a network analyser comprises a swept signal source of constant amplitude, and a receiver of constant sensitivity which is always tuned in sympathy with the instantaneous frequency of the source.

In a vector network analyser, the receiver is phase-sensitive and its output can be displayed on the instrument's display device (usually a cathode ray tube) as amplitude and/or phase against frequency (a Bode plot), or on a polar plot, or on a Smith chart. The reference for phase measurements may be the swept source's output or may be obtained from one of the accessories which are available for use with the network analyser.

A scalar analyser is similar, except that the receiver produces only amplitude information. If the unit under test produces an output frequency different from the source frequency (e.g. a mixer or frequency changer unit), there is no meaningful relation between its output phase and that of the source, so a scalar measurement is the only possible one. To cope with this type of measurement, the Anritsu vector network analyser model MS3606B (illustrated in Figure 14.4) has, among its wide range of optional accessories, an RF sensor MA4601A which permits scalar measurements to be made. This sensor is also used, in conjunction with one of a wide range of reflection coeffecient bridges, when simultaneously measuring return loss together with amplitude and either phase or group delay. Reflection (return loss) bridges are available in 50 and 75 Ω unbalanced versions covering 10 Hz to 1 GHz, and in versions for 75, 135, 150, 600 and 900 Ω balanced systems. The addition of the MH681K s-parameter test set permits the s-parameters of active devices and amplifiers to be measured. A choice of two plug-ins enables the MH681K to refer its s-parameter measurements to 50 Ω unbalanced or 75 Ω unbalanced, as required. For further information on s-parameters, see Appendix 2.

Other instruments

RF signal generators have long been fundamental items in the RF engineer's armoury and their design has advanced enormously since the days of the Marconi TF144G, known to a generation of engineers, from its wide squat shallow case, as 'the coffin'. Early types such as the TF144H were simply LC oscillators tuned by a variable capacitor in conjunction with a turret of coils for different ranges. They were designed in such a way as to minimize both the variation of output level with tuning and the amount of incidental FM which was caused when amplitude modulation was applied — and in later models fitted with a facility for frequency modulation, the amount of incidental AM caused when frequency modulation was applied. All highclass signal generators nowadays employ synthesis, so that their medium- and long-term frequency accuracy is equal to that of their ovened crystal oscillator reference. One scheme offering very low noise is direct synthesis: this technique is not to be confused with direct digital synthesis which is discussed in Chapter 8. Early synthesized signal generators using direct synthesis, such as those from General Radio, used decade synthesis whereas later generation models from Eaton/Ailtech used binary synthesis, considerably easing the design problems and resulting in a generator whose output phase noise really is nearly as good as a prime crystal oscillator. However, for reasons of economy (a direct synthesizer is complicated, and therefore expensive) most modern high-class signal generators use a VCO/PLL approach. An example of such an instrument, of advanced design, is shown in Figure 14.5. This instrument offers 0.1 Hz resolution over the complete range of 10 kHz–1.35 GHz (optionally ranging to 2.7 or 5.4 GHz) and low-phase noise. The phase noise of the companion 2040 series signal generators from the same manufacturer is even lower: –140 dBc at 10 kHz offset from carrier at 1 GHz. The very low noise of these generators is achieved using a patented development of fractional-N synthesis employing multiple accumulators, and making use of a 10 000 gate 1-micron CMOS (complentary metal-oxide-silicon) gate array ASIC (application specific integrated circuit). The ASIC also enables the implementation of a dc-coupled FM input [1]. The instrument has facilities for AM, PM and both normal and extra wideband FM.

Figure 14.5 *The 2030 series of signal generators from Marconi Instruments cover frequencies up to 5.4 GHz with 0.1 Hz resolution and +13 dBM output (+19 dBM optional). The 2040 series offers even lower phase noise.*
(Reproduced by courtesy of Marconi Instruments Ltd.)

Using the traditional approach, for tasks involving many measurements such as testing a complete radio communications system, a considerable number of different test instruments would be required. There would further be many different interconnection set-ups required during the course of testing, all of which makes this approach unattractive, especially when the test equipment has to be taken to the radios rather than vice versa. For this reason, special purpose radio communications test sets are available from a number of manufacturers. For example, the Marconi Instruments model 2955A comprises 19 different instrument functions for transceiver testing; in addition to RF and AF generators, power-, frequency- and modulation meters covering AM, FM, PM and SSB, etc., it includes DTMF (dual tone multi-frequency) and digitally-coded squelch (DCS) encoders and decoders, a digital oscilloscope function and even a POCSAG (Post Office Code Standardization Advisory Group) encoder for testing digital pagers. Consequently the 2955 series Radio Communications Test Sets can test almost any radio transceiver operating at frequencies up to 1000 MHz, including the different varieties of NMT, AMPS, TACS, Radiocom 2000 and MPT1327 networks.

The humble oscilloscope, although not normally considered as a piece of RF test gear, should not be forgotten. A conventional analogue oscilloscope, given adequate bandwidth, can be used for many RF tests. Obviously, it can be used to measure directly the peak-to-peak amplitude of a CW signal, the rms value being obtained by dividing by 2.828. This assumes that the harmonic content of the signal is low, a point which can be judged adequately if the bandwidth of the oscilloscope exceeds three times the frequency of the signal. Circuit misbehaviour, such as squegging of an oscillator, is instantly revealed by the oscilloscope where otherwise the problem might not be at all obvious.

The oscilloscope can also be used to measure the modulation index of an FM signal. Here, the oscilloscope displays a few or many cycles of the RF as required, whilst triggered from the same RF. At the left-hand side of the screen, all traces will be in phase, but moving progressively to the right, the traces will diverge to the right or left of the average, according to whether the particular trace was written when the frequency deviation was negative or positive. The point where late cycles n cycles across the screen just meet early cycles $n + 1$ cycles after the trigger point is very clearly visible; the value $n + \frac{1}{2}$ where this occurs marks the point of +/–180° peak phase deviation, from which, knowing the frequency of the modulating sinewave, the modulation index is simply derived. The oscilloscope can even be used for quite sophisticated measurements, such as eye diagrams for DPSK or similar digital modulation methods. Here, the oscilloscope displays the IF output of the transmitter modulator (or of the receiver IF) whilst it is triggered from the unmodulated IF carrier. This may be obtained from the carrier input to the modulator, or if the receiver uses synchronous demodulation, from the receiver's carrier recovery circuit. (The receiver test may be carried out with the transmitter's IF output patched into the receiver's IF strip, or alternatively it may include the RF path. In the latter case, however, either the receiver first mixer should be driven from the transmitter's final upconverter drive, or both TX and RX synthesizers should be run from the same reference.) Finally, a pulse whose frequency is that of the data clock and whose width is about 10% of the data period, is applied to the Z modulation input (bright-up input) of the oscilloscope. The pulse can be triggered by the transmitter's data clock, or obtained from the receiver's clock recovery circuit (see Figure 14.6). The bright-up pulse should have a variable delay with respect to the data clock edge: adjusting the delay to centre the pulse on the data-stable period will produce an eye diagram similar to that shown in Figure 14.2b. Note that if the transmitter modulator includes an all-pass filter providing equalization for both the transmitter and the receiver IF filtering functions, the eye diagram at the receiver's IF output should (in the absence of additive noise) be considerably cleaner and more 'open' than at the transmitter modulator's output.

Finally a word about field strength measuring equipment — used for a variety of purposes, including EMC measurements. Measuring receivers are specialized instruments which are in some respects akin to a spectrum analyser, but very different in other ways — such as not possessing a visual display. Typical examples would cover 9 kHz to 30 MHz, or 30

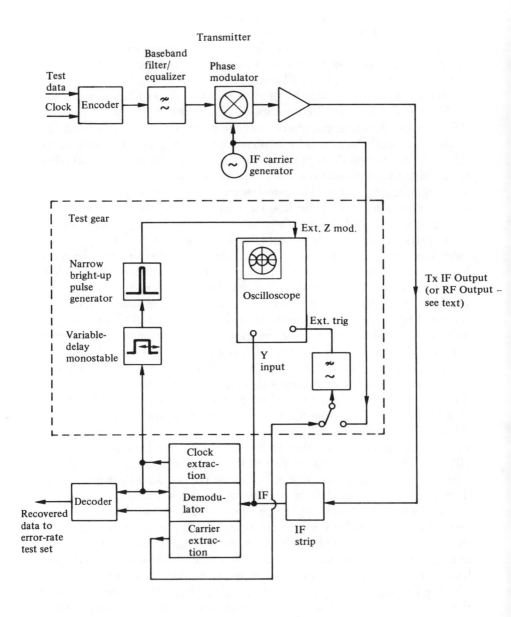

Figure 14.6   *Block diagram of digital phase-modulation radio link on test (simplified).*

MHz to 1 GHz, covering between them measurements to CISPR 16 (bands A to D). Detector response can be selected as average, peak or quasi-peak (CISPR), and in addition to spot frequency measurements, the band or any part of it can be automatically swept. The received level is output to a plotter, together a specification limit line, such as the relevant VDE limit.

Such receivers are used in conjunction with a special measuring antenna, or field probe. Simple E and H field probes have a response which, in terms of the signal strength delivered to a spectrum analyser or measuring receiver, is not constant with frequency. Nevertheless, since they are easily fabricated, they can be useful adjuncts in any RF laboratory. Figure 14.7 shows the response of simple probes in the VHF region, giving the incident field strength in terms of the measured level in dBm on, for example, a spectrum analyser, assuming the probe is in the far field of the source. More sophisticated measurement antennas cover a wide bandwidth, e.g. the HLA 6120 9 kHz–30 MHz HF Loop Antenna from Chase Electronics. This is an active antenna, providing a constant antenna factor of unity over the whole frequency range, the measured output in dBμV being numerically equal to the field strength in dBμV/m. It is ideal for the 3m magnetic field measurements to VDE 0871 and FCC

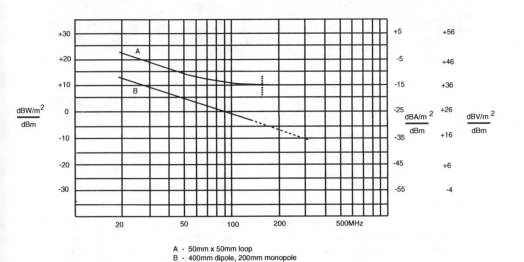

A - 50mm x 50mm loop
B - 400mm dipole, 200mm monopole

Figure 14.7 *Performance of some simple E and H field probes at VHF showing the E, H or power field strength needed to deliver 1 mW to a measuring instrument. Bear in mind that field strength measurements can seldom be relied upon to better than ±3dB.*

Figure 14.8 *The EMC20 Wideband Field Probe has an isotropic response (see text). It is shown here mounted in an anechoic chamber, with (in the background) the CBL6112 BiLog ® Antenna, which covers 30–2000 MHz.*

18. The model CBL 6112, from the same company, is in effect a compound antenna. It consists of a bi-conical (bow-tie) element and a log periodic section, permitting testing over the whole range from 30 MHz to 2 GHz with a single antenna. Primarily an emission test antenna, it will nevertheless accept powers up to 300 W for purposes of immunity testing, with field strengths up to 10 V/m or more.

The above measuring antennas are of course not isotropic, since, as was explained in Chapter 12, it is not possible to design an antenna to be isotropic. However, the EMC 20 Wideband Field Probe from Chase Electronics, covering 100 kHz to 3 GHz, is in fact isotropic. It does not infringe Maxwells equations, for the head contains three separate orthogonal sensors. The three sensors measure the electric field strength in the three axes individually, and the field strength is computed by the instrument's processor by summing the squares of the three measured values. If placed in the near field of an emitter, it measures just the E field component of the field. If placed in the far field, at at least one wavelength away and preferably three wavelengths, it again measures the E field, in volts/m, from which the H field in A/m and the power flux density in W/m^2 can be directly derived, given that the wave impedance in the far field equals that of free space, namely 377 Ω — see Figure 9 of Appendix 11.

Reference

1   Owen, D. A new approach to fractional-N synthesis. *Electronic Engineering*, 35–8 (March 1990)

Appendix 1
Useful relationships

(i) Series parallel equivalents

The following (frequency-dependent) transformation is useful where a measurement system gives the parallel components of an impedance but the series equivalent is required, or vice versa.

$Z_s = M_s \angle \phi_s$

$M_s = \sqrt{(R_s^2 + X_s^2)}$

$\phi_s = \tan^{-1} \dfrac{X_s}{R_s}$

$Z_p = M_p \angle \phi_p$

$M_p = X_p R_p / \sqrt{(R_p^2 + X_p^2)}$

$\phi_p = \tan^{-1} \dfrac{R_p}{X_p}$

$\mathcal{R} \quad \cos \phi_s = \dfrac{R_s}{\sqrt{(R_s^2 + X_s^2)}} = \dfrac{R_s}{M_s}$

$\cos \phi_p = \dfrac{X_p}{\sqrt{(R_p^2 + X_p^2)}} = \dfrac{M_p}{R_p}$

$\mathcal{I} \quad \sin \phi_s = \dfrac{X_s}{\sqrt{(R_s^2 + X_s^2)}} = \dfrac{X_s}{M_s}$

$\sin \phi_p = \dfrac{R_p}{\sqrt{(R_p^2 + X_p^2)}} = \dfrac{M_p}{X_p}$

For equivalence, $M_s = M_p$ and $\phi_s = \phi_p$

Serial to parallel:

$$R_p = \frac{R_s^2 + X_s^2}{R_s}, \qquad X_p = \frac{R_s^2 + X_s^2}{X_s}$$

Parallel to serial

$$R_s = \frac{R_p X_p^2}{R_p^2 + X_p^2}, \qquad X_s = \frac{R_p^2 X_p}{R_p^2 + X_p^2}$$

Figure A1.1

(ii) Delta/star equivalence

As in the case of (i) above, these conversions are frequency dependent.

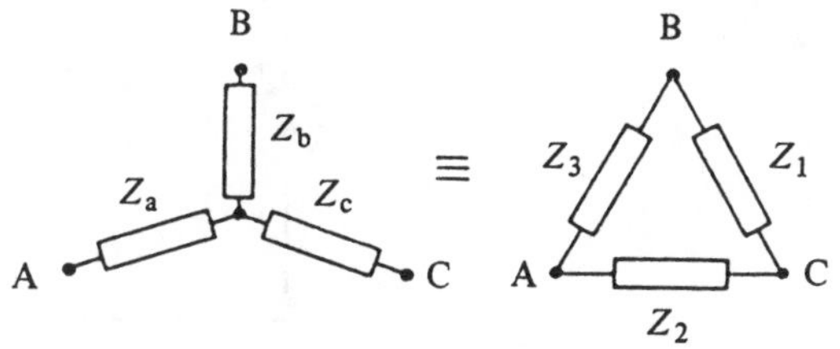

Star or wye λ Delta or mesh Δ

λ to Δ Δ to λ

$$Z_1 = Z_b + Z_c + \frac{Z_b Z_c}{Z_a} \qquad Z_a = \frac{Z_2 Z_3}{Z_1 + Z_2 + Z_3}$$

$$Z_2 = Z_a + Z_c + \frac{Z_a Z_c}{Z_b} \qquad Z_b = \frac{Z_1 Z_3}{Z_1 + Z_2 + Z_3}$$

$$Z_3 = Z_a + Z_b + \frac{Z_a Z_b}{Z_c} \qquad Z_c = \frac{Z_1 Z_2}{Z_1 + Z_2 + Z_3}$$

Figure A1.2 *The star–delta transformation (also works for impedances, enabling negative values of resistance effectively to be produced).*

(iii) Maximum power theorem

Figure A1.3 *(see p. 236). The maximum power theorem.*
a Ideal voltage source.
b Generator or source with internal resistance R_s.
c Connected to a load R_L.
d $E = 2$ V, $R_s = 1\ \Omega$. Maximum power in the load occurs when $R_L = R_s$ and $V = E/2$ (the matched condition), but only falls by 25% for $R_L = 3R_s$ and $R_L = R_s/3$. For the matched case the total power supplied by the battery is twice the power supplied to the load. On short-circuit, four times the matched load power is supplied, all dissipated internally in the battery.

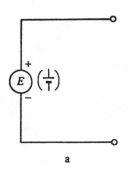

a

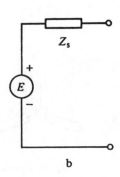

b

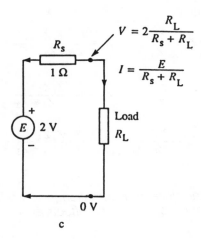

c

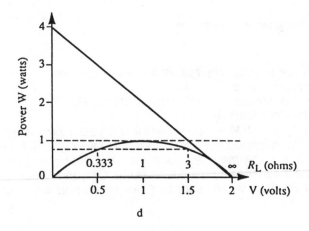

d

NOTE: Where the source impedance is not R_s but Z_s ($Z_s = M_s\angle\phi_s$) then maximum power transfer occurs when the load impedance $Z_1 = M_1\angle\phi_1 = Z_s^*$, where $Z_s^* = M_s\angle-\phi_s$. Z_s and Z_s^* are called conjugate impedances; they have the same modulus or magnitude M and the same numerical argument or phase angle ϕ, but leading in one case and lagging in the other. If the modulus of the load can be varied (e.g. by adjusting the ratio of a matching transformer) but not its phase angle, then the power transfer which can be achieved is less than the maximum (unless $\phi_1 = \phi_s$), but is at its greatest when $M_1 = M_s$.

(iv) Designing lumped component matching using the Smith chart. (Reproduced by courtesy of GEC Plessey Semiconductors Ltd.)

The main application for Smith Charts with integrated circuits is in the design of matching networks. Although these can be calculated by use of the series to parallel (and vice versa) transforms, followed by the application of Kirchoff's Laws, the method can be laborious. Although the Smith Chart as a graphical method cannot necessarily compete in terms of overall accuracy, it is nevertheless more than adequate for the majority of problems, especially when the errors inherent in practical components are taken into account.

Any impedance can be represented at a fixed frequency by a shunt conductance and susceptance (impedances as series reactance and resistance in this context). By transferring a point on the Smith Chart to a point at the same diameter but 180° away, this transformation is automatically made (see Figure A1.4) where A and B are the series and parallel equivalents.

It is often easier to change a series RC network to its equivalent parallel network for calculation purposes. This is because as a parallel network of admittances, a shunt admittance can be directly added, rather than the tortuous calculations necessary if the series form is used. Similar arguments apply to parallel networks, so in general it is best to deal with admittances for shunt components and reactances for series components.

Admittances and impedances can be easily added on the Smith Chart (see Figure A1.5). Where a series inductance is to be added to an admittance (i.e. parallel R and C), the admittance should be turned into a series impedance by the method outlined above and in Figure A1.4. The series inductance can then be added as in Figure A1.5 (see also Figure A1.6).

Point A is the starting admittance consisting of a shunt capacitance and resistance. The equivalent capacitive impedance is shown at point B. The addition of a series inductor moves the impedance to point C. The value of this inductor is defined by the length of the arc BC, and in Figure A1.6 is $-j0.5$ to $j0.43$ i.e. a total of $j0.93$. This reactance must of course be

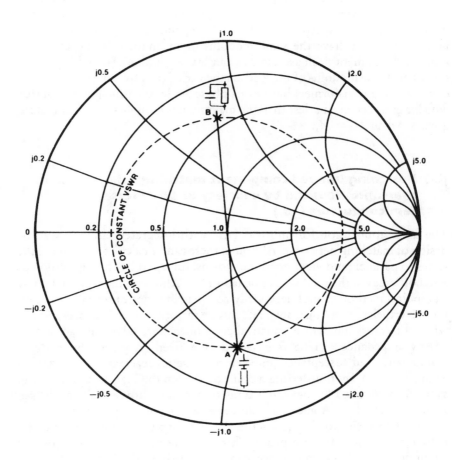

Figure A1.4 *Series reactance to parallel admittance conversion.*

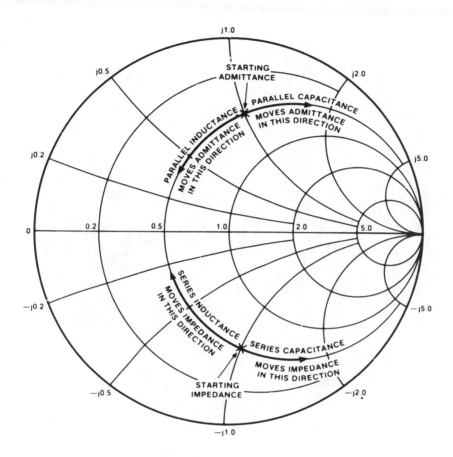

Figure A1.5 *Effects of series and shunt reactance.*

denormalized before evaluation. Point C represents an inductive impedance which is equivalent to the admittance shown at Point D. The addition of shunt reactance moves the input admittance to the centre of the chart, and has a value of $-j2.0$. Point D should be chosen such that it lies on unity impedance/conductance circle: thus a locus of points for point C exists.

This procedure allows for design of the matching at any one frequency. Wide band matching is more difficult and other techniques are needed. Of these, one of the most powerful is to absorb the reactance into a low pass filter form of ladder network: if the values are suitably chosen, the resulting input impedance is dependent upon the reflection coefficient of the filter.

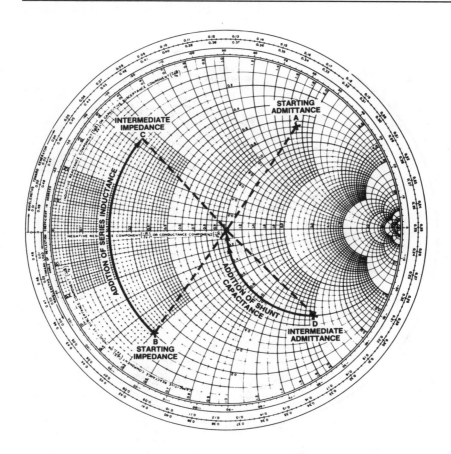

Figure A1.6   *Matching design using the Smith Chart.*

At frequencies above about 400 MHz, it becomes practical to use sections of transmission line to provide the necessary reactances, and reference to one of the standard works on the subject is recommended.*

*See Chapter 2.

Appendix 2
S-Parameters

(Reproduced by courtesy of Marconi Instruments Ltd.)

S-Parameters and Transformations

In microwave circuit design S-parameters are very useful for the full characterization of any 2 port Network.

In contrast to z, y and h-parameters, which require broadband short circuited and open circuited connections at the TEST ITEM for the measurement, S-parameters are determined with input and output terminated with the resistive characteristic impedance of test systems (generally 50 ohms in coaxial line system).

Parasitic oscillations in active devices are minimised when these devices are terminated in resistive loads.

S-parameters are complex, having a magnitude and a phase relationship, and are measured in terms of incident and reflected voltages using a VECTOR VOLTMETER.

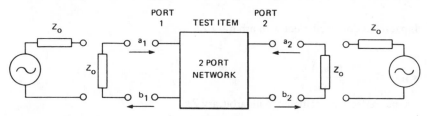

The four S-parameters are:

With Generator connected to port 1 and port 2 perfectly matched ($a_2 = o$)

INPUT-REFLECTION COEFFICIENT $\qquad S_{11} = \dfrac{b_1}{a_1}$

Looking into port 1 when port 2 is perfectly matched.

241

FORWARD-TRANSMISSION COEFFICIENT $S_{21} = \dfrac{b_2}{a_1}$

Voltage transmission coefficient from port 1 to port 2 when port 2 is perfectly matched.

With Generator connected to port 2 and port 1 perfectly matched ($a_1 = 0$)

REVERSE-TRANSMISSION COEFFICIENT $S_{12} = \dfrac{b_1}{a_2}$

Voltage transmission coefficient from port 2 to port 1 when port 1 is perfectly matched.

OUTPUT-REFLECTION COEFFICIENT $S_{22} = \dfrac{b_2}{a_2}$

Looking into port 2 when port 1 is perfectly matched.

Useful Scattering Parameters Relationships

$$b_1 = s_{11}a_1 + s_{12}a_2$$
$$b_2 = s_{21}a_1 + s_{22}a_2$$

Input reflection coefficient with arbitrary Z_L

$$s'_{11} = s_{11} + \frac{s_{12}s_{21}\Gamma_L}{1 - s_{22}\Gamma_S}$$

Output reflection coefficient with arbitrary Z_S

$$s'_{22} = s_{22} + \frac{s_{12}s_{21}\Gamma_S}{1 - s_{11}\Gamma_L}$$

Voltage gain with arbitrary Z_L and Z_S

$$A_V = \frac{V_2}{V_1} = \frac{s_{21}(1 + \Gamma_L)}{(1 - s_{22}\Gamma_L)(1 + s'_{11})}$$

$$\text{Power Gain} = \frac{\text{Power delivered to load}}{\text{Power input to network}}$$

$$\Gamma = \frac{\text{VSWR} - 1}{\text{VSWR} + 1} = \text{modulus of reflection coefficient of source or load}$$

$$G = \frac{|s_{21}|^2 (1 - |\Gamma_L|^2)}{(1 - |s_{11}|^2) + |\Gamma_L|^2 (|s_{22}|^2 - |D|^2) - 2 \operatorname{Re} (\Gamma_L N)}$$

$$D = S_{11} S_{22} - S_{12} S_{21}$$
$$M = S_{11} - D S^*_{22}$$
$$N = S_{22} - D S^*_{11}$$

Available Power Gain $= \dfrac{\text{Power available from network}}{\text{Power available from source}}$

$$G_A = \frac{|s_{21}|^2 (1 - |\Gamma_S|^2)}{(1 - |s_{22}|^2) + |\Gamma_S|^2 (|s_{11}|^2 - |D|^2) - 2 \operatorname{Re} (\Gamma_S M)}$$

Transducer Power Gain $= \dfrac{\text{Power delivered to load}}{\text{Power available from source}}$

$$G_T = \frac{|s_{21}|^2 (1 - |\Gamma_S|^2)(1 - |\Gamma_L|^2)}{|(1 - s_{11}\Gamma_S)(1 - s_{22}\Gamma_L) - s_{12}s_{21}\Gamma_L\Gamma_S|^2}$$

Unilateral Transducer Power Gain ($s_{12} = 0$)

$$G_{TU} = \frac{|s_{21}|^2 (1 - |\Gamma_S|^2)(1 - |\Gamma_L|^2)}{|1 - s_{11}\Gamma_S|^2 \, |1 - s_{22}\Gamma_L|^2}$$
$$\quad = G_0 G_1 G_2$$

$$G_0 = |s_{21}|^2$$

$$G_1 = \frac{1 - |\Gamma_S|^2}{|1 - s_{11}\Gamma_S|^2}$$

$$G_2 = \frac{1 - |\Gamma_L|^2}{|1 - s_{22}\Gamma_L|^2}$$

Maximum Unilateral Transducer Power Gain when $|s_{11}| < 1$ and $|s_{22}| < 1$

$$G_U = \frac{|s_{21}|^2}{|(1 - |s_{11}|^2)(1 - |s_{22}|)^2|}$$
$$\quad = G_0 G_{1\,\text{max}} \, G_{2\,\text{max}}$$

$$G_{i\,\text{max}} = \frac{1}{1 - |s_{ii}|^2} \qquad i = 1, 2$$

This maximum attained for $\Gamma_S = s^*_{11}$ and $\Gamma_L = s^*_{22}$

Constant Gain circles (Unilateral case: $s_{12} = 0$)

– centre of constant gain circle is on line between centre of Smith Chart and point representing s^*_{ii}

– distance of centre of circle from centre of Smith Chart:

$$r_i = \frac{g_i |s_{ii}|}{1 - |s_{ii}|^2 (1 - g_i)}$$

– radius of circle:

$$\rho_i = \frac{\sqrt{1 - g_i} \, (1 - |s_{ii}|^2)}{1 - |s_{ii}|^2 \, (1 - g_i)}$$

Where: $i = 1, 2$

and $g_i = \dfrac{G_i}{G_{i\,max}} = G_i \, (1 - |s_{ii}|^2)$

Unilateral Figure of Merit

$$u = \frac{|s_{11}s_{22}s_{12}s_{21}|}{|(1 - |s_{11}|^2) \, (1 - |s_{22}|^2)|}$$

Error Limits on Unilateral Gain Calculation

$$\frac{1}{(1 + u^2)} < \frac{G_T}{G_{TU}} < \frac{1}{(1 - u^2)}$$

Conditions for Absolute Stability

No passive source or load will cause network to oscillate if a, b, and c are all satisfied.

a. $|s_{11}| < 1, |s_{22}| < 1$

b. $\left| \dfrac{|s_{12}s_{21}| - |M^*|}{|s_{11}|^2 - |D|^2} \right| > 1$

c. $\left| \dfrac{|s_{12}s_{21}| - |N^*|}{|s_{22}|^2 - |D|^2} \right| > 1$

Condition that a two-port network can be simultaneously matched with a positve real source and load:

$K > 1$ or $C < 1$

C = Linvill C factor $= K^{-1}$

D $= s_{11}s_{22} - s_{12}s_{21}$

M $= s_{11} - Ds^*_{22}$

N $= s_{22} - Ds^*_{11}$

$$K = \frac{1 + |D|^2 - |s_{11}|^2 - |s_{22}|^2}{2 \, |s_{12}s_{21}|} = \text{Rollett Stability Factor}$$

Source and Load for Simultaneous Match

$$\Gamma_{ms} = M^* \left| \frac{B_1 \pm \sqrt{B_1^2 - 4 \, |M|^2}}{2 \, |M|^2} \right|$$

$$\Gamma_{mL} = N^* \left| \frac{B_2 \pm \sqrt{B_2^2 - 4 \, |N|^2}}{2 \, |N|^2} \right|$$

Where $B_1 = 1 + |s_{11}|^2 - |s_{22}|^2 - |D|^2$
$B_2 = 1 + |s_{22}|^2 - |s_{11}|^2 - |D|^2$

Maximum Available Power Gain, MAG
If $K > 1$,

$$MAG = \left| \frac{s_{21}}{s_{12}} (K \pm \sqrt{K^2 - 1}) \right|$$

(Use plus sign when B_1 is positive, minus sign when B_1 is negative. For definition of B_1 see 'Source and Load for Simultaneous Match', above.)

Maximum Stable Gain, MSG

$$MSG = \left| \frac{s_{21}}{s_{12}} \right|$$

Unilateral Gain – Mason

$$U = \frac{1/2 \, |(s_{21}/s_{12}) - 1|^2}{K \, |s_{21}/s_{12}| - Re(s_{21}/s_{12})}$$

Appendix 3
Attenuators (pads)

(i) Design

Designed for 1 ohm characteristic impedance

Loss D in dB	T pad		π pad		Bridged T pad	
	a	b	c	d	e	f
1	0.0575	8.668	0.1153	17.39	0.1220	8.197
2	0.1147	4.305	0.2323	8.722	0.2583	3.862
3	0.1708	2.838	0.3518	5.853	0.4117	2.427
4	0.2263	2.097	0.4770	4.418	0.5850	1.708
5	0.2800	1.645	0.6083	3.570	0.7783	1.285
6	0.3323	1.339	0.7468	3.010	0.9950	1.005
7	0.3823	1.117	0.8955	2.615	1.238	0.8083
8	0.4305	0.9458	1.057	2.323	1.512	0.6617
9	0.4762	0.8118	1.231	2.100	1.818	0.5500
10	0.5195	0.7032	1.422	1.925	2.162	0.4633
11	0.5605	0.6120	1.634	1.785	2.550	0.3912
12	0.5985	0.5362	1.865	1.672	2.982	0.3350
13	0.6342	0.4712	2.122	1.577	3.467	0.2883
14	0.6673	0.4155	2.407	1.499	4.012	0.2483
15	0.6980	0.3668	2.722	1.433	4.622	0.2167
16	0.7264	0.3238	3.076	1.377	5.310	0.1883
18	0.7764	0.2559	3.908	1.288	6.943	0.1440
20	0.8182	0.2020	4.950	1.222	9.000	0.1112
25	0.8935	0.1127	8.873	1.119	16.78	0.0597
30	0.9387	0.0633	15.81	1.065	30.62	0.0327
35	0.9650	0.0356	28.11	1.036	55.23	0.0182
40	0.9818	0.0200	50.00	1.020	99.00	0.0101
45	0.9888	0.0112	88.92	1.011	176.8	0.00567
50	0.9937	0.00633	158.1	1.0063	315.2	0.00317

(ii) Use to improve matching

(Reproduced by courtesy of Marconi Instruments Ltd)

Reduction of VSWR by Matched Attenuators

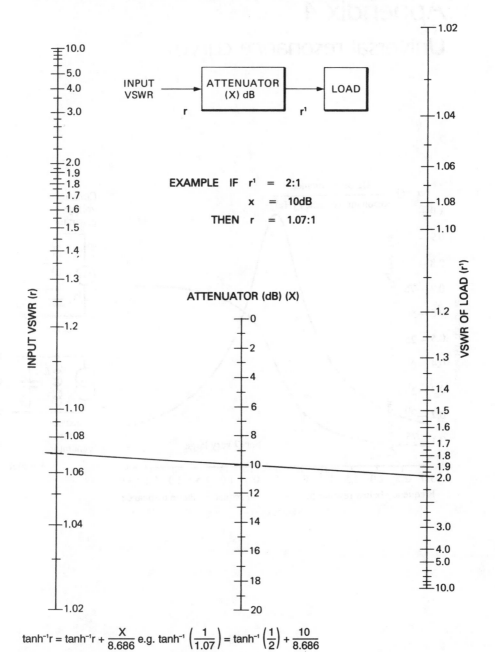

$$\tanh^{-1}r = \tanh^{-1}r + \frac{X}{8.686} \quad \text{e.g.} \ \tanh^{-1}\left(\frac{1}{1.07}\right) = \tanh^{-1}\left(\frac{1}{2}\right) + \frac{10}{8.686}$$

Appendix 4

Universal resonance curve

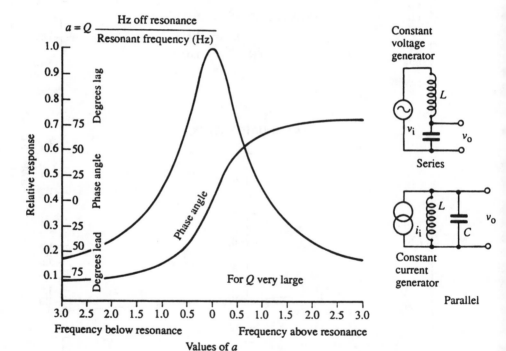

$$a = Q \, \frac{\text{Hz off resonance}}{\text{Resonant frequency (Hz)}}$$

Constant voltage generator

Series

Constant current generator

Parallel

For Q very large

Frequency below resonance   Frequency above resonance

Values of a

Appendix 5
RF cables

Data on US and UK coaxial cable types (The data in this appendix are reproduced by courtesy of Transradio Ltd.)

TRANSRADIO PART NO.	Q 98100	Q 98101	Q 98102	Q 98103	Q 98104	Q 98105	Q 98137	Q 98139	Q 98106	Q 98107	Q 98141	Q 98111	Q 98112	Q 98113	Q 98114	Q 98115	Q 98116
RG TYPE	6A/U	11A/U	22B/U	58 C/U Grey	58 C/U Black	59B/U	59 B/U Twin	59 B/U Armoured	62 A/U	62 A/U Outdoor	62 A/U Armoured	142B/U	174U	178B/U	179B/U	180B/U	188A/U
NOM. IMPEDANCE OHMS	75	75	93	50	50	75	75	75	93	93	93	50	50	50	75	95	50
NOM. CAPACITANCE pF/m	67.5	67.5	52	101	101	67.6	67.6	67.6	44.3	44.3	44.3	96.4	101.0	96.4	50.5	50.5	96.4
ATTENUATION db/100M 10MHZ	3.0	1.8	2.8	5.0	5.0	3.5	3.5	3.5	2.9	2.9	2.9	5.0	10	14	8.5	6.0	12
50MHZ	7.0	4.5	6.2	12	12	8.0	8.0	8.0	6.5	6.5	6.5	12.0	24	32	20	14	18
100MHZ	10.0	6.5	9.0	16	16	12	12	12	9.2	9.2	9.2	16	34	46	28	21	37.7
800MHZ	28	22	–	50	50	34	34	34	26	26	26	48	130	150	94	70	90
CONDUCTOR: Material	Cu W SOLID	TiC 7/0.40	2xCu 7/0.40	Cu 19/0.18	Cu 19/0.18	Cu W SOLID	Cu W SOLID	Cu W SOLID	Cu W SOLID	Cu W SOLID	Cu W SOLID	Si.Cu W SOLID	Cu W 7/0.16	Si.Cu W 7/0.10	Si.Cu W 7/0.10	Si.Cu W 7/0.10	Si.Cu W 7/0.17
DIA.MM.	0.7	1.2	1.2	0.9	0.9	0.6	0.6	0.6	0.64	0.64	0.64	0.99	0.48	0.305	0.305	0.305	0.50
DIELETRIC: Material	P.E.	P.E.	P.E.	P.E.	P.E.	P.E.	P.E.	P.E.	PE+TH	PE+TH	PE+TH	PTFE	PE	PTFE	PTFE	PTFE	PTFE
O/D(NOM.)	4.6	7.2	7.3	3.0	3.0	3.7	3.7	3.7	3.7	3.7	3.7	3.0	1.5	0.86	1.6	2.6	1.5
SCREEN: Material 1st	SiCu	Cu	TiC	TiC	TiC	Cu	Cu	Cu	Cu	Cu	Cu	Si.Cu	TiC	Si.Cu	Si.Cu	Si.Cu	Si.Cu
2nd	SiCu	–	TiC	–	–	–	–	–	–	–	–	Si.Cu	–	–	–	–	–
SHEATH: Material	PVC	PVC	PVC	PVC	PVC	PVC	PVC	PVC	PVC	PE	PVC	FEP	PVC	FEP	FEP	FEP	PTFE
O/D(NOM.)	8.4	10.3	10.3	4.9	4.9	6.2	6.2	–	6.2	6.2	–	4.9	2.54	1.9	2.54	3.7	2.8
Weight Approx KG/KM	119	143	180	43	43	48	96	–	56	57	–	74	11.8	7.4	14.8	28.1	16.2
MIN. BENDING RADIUS	102	114	51	51	51	51	–	–	51	116	–	51	25.4	25.4	25.4	50.8	25.4

TRANSRADIO PART NO.	Ω 98117	Ω 98119	Ω 98120	Ω 98126	Ω 98122	Ω 98123	Ω 98124	Ω 98127
RG TYPE	196A/U	213U	214U	215U	217U	218U	223U	316U
NOM. IMPEDANCE OHMS	50	50	50	50	50	50	50	50
NOM. CAPACITANCE pF/m	96.4	101.0	101.0	101.0	101.0	101.0	101.0	96.4
ATTENUATION db/100m 10MHZ	22	1.9	2.4	2	1.9	0.7	5.0	12
50MHZ	28.0	4.6	5.8	4.9	4.4	1.8	12.0	18
100MHZ	47.2	6.8	7.2	8.8	6.2	2.7	17	37.7
800MHZ	134	23	28	23	19	9.4	4.8	90
CONDUCTOR: Material	Si.Cu.W 7/0.10	Cu. 7/0.75	Si.Cu. 7/0.75	Cu. 7/0.75	Cu. SOLID	Cu. SOLID	Si.Cu. SOLID	Si.Cu.W 7/0.17
DIA.MM	0.305	2.2	2.2	2.2	2.7	4.9	0.89	0.50
DIELECTRIC: Material	PTFE	PE	PE	PE	PE	PE	PE	PTFE
O/D(NOM.)	0.86	7.3	7.3	7.3	9.4	17.3	2.9	1.5
SCREEN: Material 1st	Si.Cu.	Cu.	Si.Cu.	Cu.	Cu.	Cu.	Si.Cu.	Si.Cu.
2nd	—	—	Si.Cu.	—	Cu.	—	Si.Cu.	—
SHEATH: Material	PTFE	PVC	PVC	PVCA	PVC	PVC	PVC	FEP
O/D(NOM.)	2.0	10.3	10.7	12.1	13.8	22	5.5	2.6
Weight: Approx KG/KM	8.8	146	186	225	297	680	50.3	17.8
MIN BENDING RADIUS	25.4	114	127	152	197	254	51	25.4

TRANSRADIO PART NO.	Q 98186	Q 98187	Q 98188	Q 98189	Q 98185	Q 98190	Q 98193	Q 98192
URM TYPE	43	57	67	70	74	76	90	96
NOM. IMPEDANCE OHMS	50	75	50	75	50	50	75	96
NOM. CAPACITANCE pF/m	95	68	100	67	100	100	67	40
ATTENUATION db/100m 100MHZ	13.0	6.1	6.8	15.2	3.2	15.5	11.2	7.9
200MHZ	18.5	9.0	9.9	21.8	4.8	22.2	16.1	11.2
300MHZ	23.0	11.5	12.5	27.0	6.1	27.4	20.0	13.8
600MHZ	34.0	17.0	18.5	39.1	9.6	39.8	29.3	19.7
1000MHZ	45.0	23.0	25.0	51.7	13.7	52.7	39.1	25.8
CONDUCTOR: Material	Cu. SOLID	Cu. SOLID	Cu. 7/0.77	Cu. 7/0.19	Cu. SOLID	Cu. 7/0.32	Cu.W. SOLID	Cu.W. SOLID
DIA.MM.	0.90	1.15	—	—	5.0	—	0.60	0.64
DIELECTRIC: Material	PE	PE	PE	PE	PE	PE	PE	S A S P E
O/D(NOM.)	2.95	7.25	7.25	3.25	17.30	2.95	3.70	3.70
SCREEN: Material 1st	Cu.	Cu.	Cu.	Cu.	Cu.	Cu.	Cu.	Cu.
2nd	—	—	—	—	—	—	—	—
SHEATH: Material	PVC	PVC	PVC	PVC	PVC	PVC	PVC	PVC
O/D(NOM.)	5.0	10.3	10.3	5.8	22.0	5.0	6.0	6.0
Weight: Approx KG/KM	42	154	157	45	690	39	66	42
MIN. BENDING RADIUS	25	50	50	30	110	25	30	30

Appendix 6
Wire gauges and related information

Nominal diameter (mm)	Tolerance	Enamelled diameter Grade 1		Enamelled diameter Grade 2		Nom. resistance Ohms m at 20°c	Weight (kg/km)	Nominal diameter (mm)
		Min.	Max.	Min.	Max.			
0.032	±0.0015	0.035	0.040	0.035	0.043	21.44	0.0072	0.032
0.036	±0.0015	0.040	0.045	0.041	0.049	16.94	0.0091	0.036
0.040	±0.002	0.044	0.050	0.047	0.054	13.72	0.0112	0.040
0.045	±0.002	0.050	0.056	0.054	0.061	10.84	0.0142	0.045
0.050	±0.002	0.056	0.062	0.060	0.068	8.781	0.0175	0.050
0.056	±0.002	0.062	0.069	0.066	0.076	7.000	0.0219	0.056
0.063	±0.002	0.068	0.078	0.076	0.085	5.531	0.0277	0.063
0.071	±0.003	0.076	0.088	0.086	0.095	4.355	0.0352	0.071
0.080	±0.003	0.088	0.098	0.095	0.105	3.430	0.0447	0.080
0.090	±0.003	0.098	0.110	0.107	0.117	2.710	0.0566	0.090
0.100	±0.003	0.109	0.121	0.119	0.129	2.195	0.0699	0.100
0.112	±0.003	0.122	0.134	0.130	0.143	1.750	0.0877	0.112
0.125	±0.003	0.135	0.149	0.146	0.159	1.405	0.109	0.125
0.132	±0.003	0.143	0.157	0.153	0.165	1.260	0.122	0.132
0.140	±0.003	0.152	0.166	0.164	0.176	1.120	0.137	0.140
0.150	±0.003	0.163	0.177	0.174	0.187	0.9757	0.157	0.150
0.160	±0.003	0.173	0.187	0.187	0.199	0.8575	0.179	0.160
0.170	±0.003	0.184	0.198	0.197	0.210	0.7596	0.202	0.170
0.180	±0.003	0.195	0.209	0.209	0.222	0.6775	0.226	0.180
0.190	±0.003	0.204	0.220	0.219	0.233	0.6081	0.252	0.190
0.200	±0.003	0.216	0.230	0.232	0.245	0.5488	0.280	0.200
0.212	±0.003	0.229	0.243	0.247	0.260	0.4884	0.314	0.212
0.224	±0.003	0.240	0.256	0.258	0.272	0.4375	0.351	0.224
0.236	±0.003	0.252	0.268	0.268	0.285	0.3941	0.389	0.236
0.250	±0.004	0.267	0.284	0.284	0.301	0.3512	0.437	0.250
0.265	±0.004	0.282	0.299	0.299	0.317	0.3126	0.491	0.265
0.280	±0.004	0.298	0.315	0.315	0.334	0.2800	0.548	0.280
0.300	±0.004	0.319	0.336	0.336	0.355	0.2439	0.629	0.300
0.315	±0.004	0.334	0.352	0.353	0.371	0.2212	0.694	0.315
0.335	±0.004	0.355	0.374	0.374	0.392	0.1956	0.784	0.335
0.355	±0.004	0.375	0.395	0.395	0.414	0.1742	0.881	0.355
0.375	±0.004	0.395	0.416	0.416	0.436	0.1561	0.983	0.375
0.400	±0.005	0.421	0.442	0.442	0.462	0.1372	1.12	0.400
0.425	±0.005	0.447	0.468	0.468	0.489	0.1215	1.26	0.425
0.450	±0.005	0.472	0.495	0.495	0.516	0.1084	1.42	0.450
0.475	±0.005	0.498	0.522	0.521	0.544	0.09730	1.58	0.475
0.500	±0.005	0.524	0.547	0.547	0.569	0.08781	1.75	0.500
0.530	±0.006	0.555	0.580	0.579	0.602	0.07814	1.96	0.530
0.560	±0.006	0.585	0.610	0.610	0.632	0.070.00	2.19	0.560
0.600	±0.006	0.625	0.652	0.650	0.674	0.06098	2.52	0.600
0.630	±0.006	0.657	0.684	0.683	0.706	0.05531	2.77	0.630
0.670	±0.007	0.698	0.726	0.726	0.748	0.04890	3.14	0.670
0.710	±0.007	0.738	0.767	0.766	0.790	0.04355	3.52	0.710
0.750	±0.008	0.779	0.809	0.808	0.832	0.03903	3.93	0.750
0.800	±0.008	0.830	0.861	0.860	0.885	0.03430	4.47	0.800
0.850	±0.009	0.881	0.913	0.912	0.937	0.03038	5.05	0.850
0.900	±0.009	0.932	0.965	0.964	0.990	0.02710	5.66	0.900
0.950	±0.010	0.983	1.017	1.015	1.041	0.02432	6.31	0.950
1.00	±0.010	1.034	1.067	1.067	1.093	0.02195	6.99	1.00
1.06	±0.011	1.090	1.130	1.123	1.155	0.01954	7.85	1.06
1.12	±0.011	1.150	1.192	1.181	1.217	0.01750	8.77	1.12
1.18	±0.012	1.210	1.254	1.241	1.279	0.01577	9.73	1.18
1.25	±0.013	1.281	1.325	1.313	1.351	0.01405	10.9	1.25
1.32	±0.013	1.351	1.397	1.385	1.423	0.01260	12.2	1.32
1.40	±0.014	1.433	1.479	1.466	1.506	0.01120	13.7	1.40
1.50	±0.015	1.533	1.581	1.568	1.608	0.009757	15.7	1.50
1.60	±0.016	1.633	1.683	1.669	1.711	0.008575	17.9	1.60
1.70	±0.017	1.733	1.785	1.771	1.813	0.007596	20.2	1.70
1.80	±0.018	1.832	1.888	1.870	1.916	0.006775	22.7	1.80
1.90	±0.019	1.932	1.990	1.972	2.018	0.006081	25.2	1.90
2.00	±0.020	2.032	2.092	2.074	2.120	0.005488	28.0	2.00

Manufacturers offer several grades of insulation material and thickness. The thicker coatings are recommended for high-voltage transformer applications. The most popular coating materials are 'self-fluxing', i.e. do not require a separate end stripping operation before soldering.

No.	SWG		BWG		AWG or B & S		No.	SWG		BWG		AWG or B & S	
	in	mm	in	mm	in	mm		in	mm	in	mm	in	mm
4/0	0.400	10.160	0.454	11.532	0.4600	11.684	24	0.022	0.559	0.022	0.559	0.0201	0.511
3/0	0.372	9.449	0.425	10.795	0.4096	10.404	25	0.020	0.508	0.020	0.508	0.0179	0.455
2/0	0.348	8.839	0.380	9.652	0.3648	9.266	26	0.018	0.457	0.018	0.457	0.0159	0.404
0	0.324	8.230	0.340	8.636	0.3249	8.252	27	0.0164	0.417	0.016	0.406	0.0142	0.361
1	0.300	7.620	0.300	7.620	0.2893	7.348	28	0.0148	0.376	0.014	0.356	0.0126	0.320
2	0.276	7.010	0.284	7.214	0.2576	6.543	29	0.0136	0.345	0.013	0.330	0.0113	0.287
3	0.252	6.401	0.259	6.579	0.2294	5.827	30	0.0124	0.315	0.012	0.305	0.0100	0.254
4	0.232	5.893	0.238	6.045	0.2043	5.189	31	0.0116	0.295	0.010	0.254	0.0089	0.226
5	0.212	5.385	0.220	5.588	0.1819	4.620	32	0.0108	0.274	0.009	0.229	0.0080	0.203
6	0.192	4.877	0.203	5.156	0.1620	4.115	33	0.0100	0.254	0.008	0.203	0.0071	0.180
7	0.176	4.470	0.180	4.572	0.1443	3.665	34	0.0092	0.234	0.007	0.178	0.0063	0.160
8	0.160	4.064	0.165	4.191	0.1285	3.264	35	0.0084	0.213	0.005	0.127	0.0056	0.142
9	0.144	3.658	0.148	3.759	0.1144	2.906	36	0.0076	0.193	0.004	0.102	0.0050	0.127
10	0.128	3.251	0.134	3.404	0.1019	2.588	37	0.0068	0.173			0.0045	0.114
11	0.116	2.946	0.120	3.048	0.0907	2.304	38	0.0060	0.152			0.0040	0.102
12	0.104	2.642	0.109	2.769	0.0808	2.052	39	0.0052	0.132			0.0035	0.090
13	0.092	2.337	0.095	2.413	0.0720	1.829	40	0.0048	0.122			0.0031	0.079
14	0.080	2.032	0.083	2.108	0.0641	1.628	41	0.0044	0.112			0.0028	0.071
15	0.072	1.829	0.072	1.829	0.0571	1.450	42	0.0040	0.102			0.0025	0.063
16	0.064	1.626	0.065	1.651	0.0508	1.290	43	0.0036	0.091			0.0022	0.056
17	0.056	1.422	0.058	1.473	0.0453	1.151	44	0.0032	0.081			0.0020	0.051
18	0.048	1.219	0.049	1.245	0.0403	1.024	45	0.0028	0.071			0.00176	0.045
19	0.040	1.016	0.042	1.067	0.0359	0.912	46	0.0024	0.061			0.00157	0.040
20	0.036	0.914	0.035	0.889	0.0320	0.813	47	0.0020	0.051			0.00140	0.036
21	0.032	0.813	0.032	0.813	0.0285	0.724	48	0.0016	0.041			0.00124	0.031
22	0.028	0.711	0.028	0.711	0.0253	0.643	49	0.0012	0.030			0.00111	0.028
23	0.024	0.610	0.025	0.635	0.0226	0.574	50	0.0010	0.025			0.00099	0.025

Appendix 7
Ferrite manufacturers

The following is a representative list of companies active in the USA and UK, from the large number of manufacturers of ferrites. It is included by way of illustration only and does not claim to be exhaustive. No responsibility can be taken for the accuracy of the details given. Many of the companies listed have subsidiaries or agents in most major countries of the developed world. In some cases, an entry is itself the national subsidiary of a company based in another country.

- Fair-Rite Products Corporation, PO Box J, Commercial Row, Wallkill, New York 12589, USA; Tel. (914) 895-2055. UK agent: Dexter Magnetic Materials; Tel. 01753 680011
- Ferroperm UK Ltd., Vauxhall Industrial Estate, Ruabon, Wrexham, Clwyd LL14 6HY, UK; Tel. 01987 823990
- Ferroxcube, Amperex Electronics Corporation 5083 Kings Highway, Saugerties, New York 12477, USA; Tel. (914) 246-2811
- Indiana General, 1168 Barranca Drive, El Paso, Texas 79935, USA; Tel. (915) 593-1621
- Iskra Electronics Incorporated, 222 Sherwood Avenue, Farmingdale, New York 11735, USA; Tel. (516) 753-0400
- Iskra Ltd., Components Group, Redlands, Coulsdon, Surrey CR33 2HT, UK; Tel. 0181 668 7141
- Krystinel Corp., 126 Pennsylvania Avenue, Paterson, New Jersey 07509, USA; Tel. (201) 345-8900
- Neosid Ltd., Icknield Way West, Letchworth, Herts SG6 4AS, UK; Tel. 01706 481000. US Agent: MMg Distribution Co. Inc.; Tel. (201) 389 4411 (East), (619) 591 4773 (West)
- Philips Components Ltd., Mullard House, Torrington Place, London WC1 7HD, UK; Tel. 0171 580 6633

- SEI Ltd., Times Mill, Dawson Street, Heywood, Lancashire OL10 4NE, UK; Tel. 01706 67501
- Siemens plc., Passive Components, Siemens House, Windmill Road, Sunbury-on-Thames, Middlesex TW16 7HS, UK; Tel. 01932 785691

Appendix 8
Types of modulation — classification

Old and new designations of emissions

Classification (based on old method)

Type of modulation of main carrier	Type of transmission	Additional characteristics	Previous designation	New designation
Amplitude modulation	With no modulation	—	A0	NON
	Telegraphy			
	Morse telegraphy	—	A1	A1A
	Teletype telegraphy	—	A1	A1B
	Morse tel., sound-mod.	—	A2	A2A
	Teletype telegraphy	—	A2	A2B
	Morse telegraphy	SSB, suppressed carrier	A2J	J2A
	Teletype telegraphy	suppressed carrier	A2J	J2B
	Morse telegraphy	reduced carrier	A2A	R2A
	Morse telegraphy	full carrier	A2H	H2A
		f. autom. reception	A2H	H2B
	Telephony	DSB	A3	A3E
		SSB, reduced carrier	A3A	R3E
		full carrier	A3H	H3E
		suppressed carrier	A3J	J3E
		Two independent sidebands	A3B	B8E
	Facsimile	—	A4	A3C
		SSB, reduced carrier	A4A	R3C
		suppressed carrier	A4J	J3C
	Television (video)	DSB	A5	A3F
		Vestigial sideband	A5C	C3F
		SSB, suppressed carrier	A5J	J3F
	Multichannel voice-frequency telegraphy	SSB, reduced carrier	A7A	R7B
		suppressed carrier	A7J	J7B
	Cases not covered by the above			
		—	A9	AXX
		DSB, 1 channel, with quantized or digital information		
		without mod. subcarrier	A9	A1D
		with mod. subcarrier	A9	A2D
		Two independent sidebands	A9B	B9W
	Morse telegraphy	SSB, suppr. carrier 1 channel, with quantized or digital information		
		with mod. subcarrier	A9J	J2A
	Teletype telegraphy	As above	A9J	J2B
	Telecommand	As above	A9J	J2D
Frequency modulation (or phase modulation)	Telegraphy by frequency-shift keying without modulating audio frequency			
	Morse telegraphy	—	F1	F1A
	Teletype telegraphy	—	F1	F1B
	Telegraphy by on-off keying of			

	frequency modulating audio frequency			
	Morse telegraphy	—	F2	F2A
	Teletype telegraphy	—	F2	F2B
	Telephony and sound broadcasting	—	F3	F3E
		Phase modulation, VHF–UHF radiotelephony	F3	G3E
	Facsimile	1 channel, with analog inform.	F4	F3C
		with quantized or digital information		F1C
		without mod. subcarr.	F4	F2C
		with mod. subcarrier	F4	
	Television (video	—	F5	F3F
	Four-frequency diplex telegraphy	—	F6	F7B
	Cases not covered by the above	—	F9	FXX
	Telecommand	1 channel, with quantized or digital information		
		without mod. subcarr.	F9	F1D
		with mod. subcarrier	F9	F2D
Pulse modulation	Pulsed carrier without any modulation (e.g. radar)	—	P0	PON
	Telegraphy	—	P1D	K1A
		Modulation of pulse amplitude	P2D	K2A
		]pulse duration	P2E	L2A
		pulse phase	P2F	M2A
	Telephony	Modulation of pulse amplitude	P3D	K2E
		pulse duration	P3E	L3E
		pulse phase	P3G	V3E
	Cases not covered by the above with pulse-modulated main carrier	—	P9	XXX

Example: $\dfrac{\text{2K70 J3E **}}{1 \quad 2 \quad 3}$ = SSB Telephony, suppressed carrier, bandwidth 2700 Hz

1. Three digits plus H.K. M or G (Hz, kHz, MHz or GHz) occupying decimal point place — necessary bandwidth.
2. Three characters (per table above) indicating type of emission.
3. Two optional characters giving further information on type of transmission.

Appendix 9
Quartz crystals

(Reproduced by courtesy of SEI Ltd, a GEC company.)

The properties of a quartz crystal operating near to a frequency of resonance can be represented by an equivalent circuit consisting of an inductance (L_1) a capacitance (C_1) and a resistance (R_1), shunted by second capacitance (C_0). The elements L_1, C_1 and R_1 have no physical existence and are introduced to provide an electrical model of a vibrating crystal plate. The commonly used simplified equivalent circuit is shown as Figure 1.

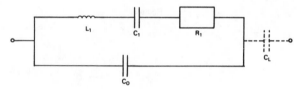

Figure 1.

The L_1, C_1, R_1 branch is known as the motional arm where L_1 is a function of the vibrating mass, C_1 represents the compliance and R_1 represents the sum of the crystal losses. C_0 is the sum of the capacitance between the crystal electrodes plus the capacitance introduced by the crystal terminals and the metal enclosure.

The crystal impedance varies rapidly in the immediate vicinity of the crystal resonance frequencies as shown in Figure 2. There are two zero phase frequencies, one at series resonance (f_S) and one at parallel or anti-resonance (f_a).

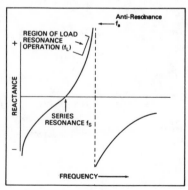

Figure 2.

Series Resonance. When a crystal is operating at series resonance its impedance at f_s is near to zero but a low active resistance remains which is known as the equivalent series resistance (ESR). The ESR value (expressed in ohms) is a measure of crystal activity and is used as an acceptance criterion.

Parallel or Anti-Resonance. When a crystal is operating at parallel resonance its impedance reaches its peak at f_a, as shown in Figure 2. Often the load circuit causes the reactive impedance to resonate in parallel or in series with the oscillator's load capacitance C_L. When a crystal is operating in this condition (f_L) the value of C_L should be precisely specified and to avoid instability the value of the load capacitance should be several times greater than the value of C_O. (Typical range of values for $C_L = 20\,pF$ to $60\,pF$.)

The frequency temperature characteristics of AT-Cut high frequency crystals show a cubic characteristic which, dependent upon the crystal plate design or mode of vibration, has an inflexion point which may be between $+27^\circ C$ and $+31^\circ C$. By careful control of the crystal cutting angle the two turning points of the curve can be positioned to provide a minimum total deviation of the crystal frequency over a specified temperature range. The frequency/temperature characteristics for the AT-Cut, shown in Figure 3, are substantially valid for most fundamental and overtone types.

Typical frequency/temperature variations

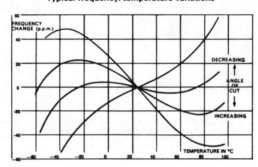

Figure 3.

Appendix 10
Elliptic filters

The following small subset of tables with their schematics are reprinted with permission from 'On the Design of Filters by Synthesis' by R. Saal and E. Ulbricht, *IRE Transactions on Circuit Theory*, December 1958, pp. 284–328. (© 1958 IRE (now IEEE)). The tables are normalized to $f = 1$ rad/s = $1/(2\pi)$ Hz, $Z_0 = 1$ Ω, L in henrys, C in farads.

(Note: In using the following tables with the schematics, for example, schematic a on page 262 corresponds with the top line of column headings of Tables A10.1–3. Similarly, schematic b corresponds with the bottom line of column headings of the tables.)

The original gives designs for filters up to the eleventh order. Designs are presented here for third and fifth order filters with 1 dB, 0.5 dB and 0.1 dB pass-band ripples, and for sixth, seventh and ninth order 0.18 dB ripple filters. For the 6-pole case, two designs are given. One is the basic 6-pole version designed to work from a normalized source impedance of unity into a normalized load impedance of 0.667 (or 1.5 for the T section design). This results in a 0.18 dB insertion loss at dc, due to the 1.5:1 VSWR. The other is a version designed to work between normalized impedances of unity at both ends and consequently has a zero pass-band loss at dc similar to that of a 5-pole filter. The first version offers a slightly faster cut-off in the stop band and is therefore to be preferred, provided that the different terminating impedances can be conveniently accommodated.

3 pole

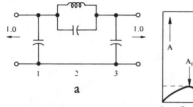 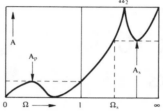 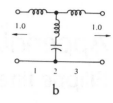

a

b

Table A10.1 $A_p = 1$ dB

Ω_s	A_s [db]	C_1	C_2	L_2	Ω_2	C_3
1.295	20	1.570	0.805	0.613	1.424	1.570
1.484	25	1.688	0.497	0.729	1.660	1.688
1.732	30	1.783	0.322	0.812	1.954	1.783
2.048	35	1.852	0.214	0.865	2.324	1.852
2.418	40	1.910	0.145	0.905	2.762	1.910
2.856	45	1.965	0.101	0.929	3.279	1.965
Ω_s	A_s [db]	L_1	L_2	C_2	Ω_2	L_3

(© 1958 IRE (now IEEE))

Table A10.2 $A_p = 0.5$ dB

Ω_s	A_s [db]	C_1	C_2	L_2	Ω_2	C_3
1.416	20	1.267	0.536	0.748	1.578	1.267
1.636	25	1.361	0.344	0.853	1.846	1.361
1.935	30	1.425	0.226	0.924	2.189	1.425
2.283	35	1.479	0.152	0.976	2.600	1.479
2.713	40	1.514	0.102	1.015	3.108	1.514
Ω_s	A_s [db]	L_1	L_2	C_2	Ω_2	L_3

(© 1958 IRE (now IEEE))

Table A10.3 $A_p = 0.1$ dB

Ω_s	A_s [db]	C_1	C_2	L_2	Ω_2	C_3
1.756	20	0.850	0.290	0.871	1.986	0.850
2.082	25	0.902	0.188	0.951	2.362	0.902
2.465	30	0.941	0.125	1.012	2.813	0.941
2.921	35	0.958	.0837	1.057	3.362	0.958
3.542	40	0.988	.0570	1.081	4.027	0.988
Ω_s	A_s [db]	L_1	L_2	C_2	Ω_2	L_3

(© 1958 IRE (now IEEE))

5 pole

Table A10.4 $A_p = 1$ dB

Ω_s	A_s [db]	C_1	C_2	L_2	Ω_2	C_3	C_4	L_4	Ω_4	C_5
1.145	35	1.783	0.474	0.827	1.597	1.978	1.487	0.488	1.174	1.276
1.217	40	1.861	0.372	0.873	1.755	2.142	1.107	0.578	1.250	1.427
1.245	45	1.923	0.293	0.947	1.898	2.296	0.848	0.684	1.313	1.553
1.407	50	1.933	0.223	0.963	2.158	2.392	0.626	0.750	1.459	1.635
1.528	55	1.976	0.178	0.986	2.387	2.519	0.487	0.811	1.591	1.732
1.674	60	2.007	0.141	1.003	2.660	2.620	0.380	0.862	1.747	1.807
1.841	65	2.036	0.113	1.016	2.952	2.703	0.301	0.901	1.920	1.873
2.036	70	2.056	.0890	1.028	3.306	2.732	0.239	0.934	2.117	1.928
Ω_s	A_s [db]	L_1	L_2	C_2	Ω_2	L_3	L_4	C_4	Ω_4	L_5

Table A10.5 $A_p = 0.5$ dB

Ω_s	A_s [db]	C_1	C_2	L_2	Ω_2	C_3	C_4	L_4	Ω_4	C_5
1.186	35	1.439	0.358	0.967	1.700	1.762	1.116	0.600	1.222	1.026
1.270	40	1.495	0.279	1.016	1.878	1.880	0.840	0.696	1.308	1.114
1.369	45	1.530	0.218	1.063	2.077	1.997	0.627	0.795	1.416	1.241
1.481	50	1.563	0.172	1.099	2.300	2.113	0.482	0.875	1.540	1.320
1.618	55	1.559	0.134	1.140	2.558	2.188	0.369	0.949	1.690	1.342
1.782	60	1.603	0.108	1.143	2.847	2.248	0.291	0.995	1.858	1.449
1.963	65	1.626	.0860	1.158	3.169	2.306	0.230	1.037	2.048	1.501
2.164	70	1.624	.0679	1.178	3.536	2.319	0.182	1.078	2.258	1.521
Ω_s	A_s [db]	L_1	L_2	C_2	Ω_2	L_3	L_4	C_4	Ω_4	L_5

Table A10.6 $A_p = 0.1$ dB

Ω_s	A_s [db]	C_1	C_2	L_2	Ω_2	C_3	C_4	L_4	Ω_4	C_5
1.309	35	0.977	0.230	1.139	1.954	1.488	0.742	0.740	1.350	0.701
1.414	40	1.010	0.177	1.193	2.176	1.586	0.530	0.875	1.468	0.766
1.540	45	1.032	0.140	1.228	2.412	1.657	0.401	0.968	1.605	0.836
1.690	50	1.044	0.1178	1.180	2.682	1.726	0.283	1.134	1.765	0.885
1.860	55	1.072	0.0880	1.275	2.985	1.761	0.241	1.100	1.942	0.943
2.048	60	1.095	0.0699	1.292	3.328	1.801	0.192	1.148	2.130	0.988
2.262	65	1.108	0.0555	1.308	3.712	1.834	0.151	1.191	2.358	1.022
2.512	70	1.112	0.0440	1.319	4.151	1.858	0.119	1.225	2.619	1.044
Ω_s	A_s [db]	L_1	L_2	C_2	Ω_2	L_3	L_4	C_4	Ω_4	L_5

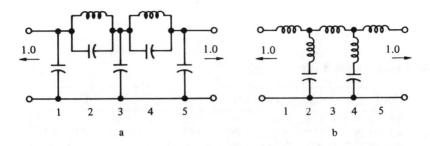

a   b

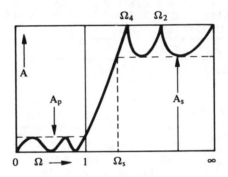

6 pole Loss = A_p at 0 Hz

Table A10.7 A_p = 0.18 dB

Ω_s	A_s[db]	C_1	C_2	L_2	Ω_2	C_3	C_4	L_4	Ω_4	C_5	L_6
3.751 039	112.5	1.299	0.0250	1.344	5.452 491	2.142	0.0468	1.412	3.888 329	2.017	0.8828
3.535 748	109.3	1.296	0.0283	1.341	5.133 037	2.135	0.0530	1.405	3.664 543	2.012	0.8830
3.344 698	106.3	1.293	0.0318	1.337	4.849 152	2.126	0.0596	1.397	3.465 915	2.006	0.8831
3.174 064	103.4	1.290	0.0355	1.333	4.595 218	2.118	0.0666	1.389	3.288 476	2.000	0.8833
3.020 785	100.7	1.286	0.0395	1.328	4.366 743	2.108	0.0740	1.380	3.120 050	1.993	0.8835
2.882 384	98.1	1.283	0.0436	1.324	4.160 091	2.009	0.0818	1.371	2.985 065	1.987	0.8837
2.756 834	95.6	1.279	0.0480	1.319	3.972 284	2.089	0.0901	1.362	2.854 418	1.979	0.8839
2.642 462	93.3	1.275	0.0527	1.314	3.800 865	2.078	0.0989	1.352	2.735 370	1.972	0.8841
2.537 873	91.0	1.270	0.0576	1.309	3.643 786	2.067	0.1081	1.341	2.626 475	1.964	0.8843
2.441 895	88.8	1.266	0.0627	1.303	3.499 325	2.055	0.1177	1.331	2.526 516	1.956	0.8845
2.353 536	86.7	1.261	0.0680	1.297	3.366 027	2.043	0.1279	1.320	2.434 463	1.948	0.8848
2.271 953	84.6	1.256	0.0736	1.291	3.242 651	2.031	0.1385	1.308	2.349 441	1.939	0.8850
2.196 422	82.6	1.251	0.0795	1.285	3.128 134	2.018	0.1497	1.296	2.270 699	1.930	0.8853
2.126 320	80.7	1.246	0.0857	1.279	3.021 559	2.005	0.1613	1.284	2.197 588	1.921	0.8855
2.061 103	78.0	1.240	0.0921	1.272	2.922 132	1.991	0.1735	1.271	2.120 540	1.911	0.8858
2.000 308	77.1	1.235	0.0988	1.265	2.829 162	1.977	0.1863	1.257	2.066 092	1.901	0.8861
1.943 517	75.3	1.220	0.1057	1.258	2.742 042	1.962	0.1996	1.244	2.006 790	1.891	0.8864
1.890 370	73.6	1.223	0.1130	1.250	2.660 241	1.947	0.2136	1.230	1.951 268	1.881	0.8867
1.840 548	72.0	1.216	0.1206	1.243	2.583 290	1.931	0.2281	1.215	1.899 195	1.870	0.8870
1.793 769	70.4	1.210	0.1285	1.235	2.510 772	1.915	0.2433	1.200	1.850 277	1.859	0.8873
1.749 781	68.8	1.203	0.1367	1.226	2.442 318	1.899	0.2592	1.185	1.804 254	1.847	0.8877
1.708 362	67.3	1.196	0.1452	1.218	2.377 598	1.882	0.2758	1.169	1.760 893	1.835	0.8880
1.669 312	65.8	1.189	0.1541	1.209	2.316 318	1.864	0.2931	1.153	1.719 987	1.823	0.8884
1.632 615	64.3	1.181	0.1634	1.200	2.258 212	1.847	0.3112	1.137	1.681 350	1.811	0.8887
1.597 615	62.8	1.174	0.1730	1.191	2.203 043	1.828	0.3301	1.120	1.644 814	1.798	0.8891
1.564 602	61.4	1.166	0.1830	1.181	2.150 505	1.810	0.3498	1.103	1.610 227	1.785	0.8895
1.533 460	60.0	1.158	0.1934	1.172	2.100 673	1.791	0.3704	1.085	1.577 454	1.771	0.8898
1.503 888	58.7	1.149	0.2043	1.161	2.053 102	1.771	0.3920	1.067	1.546 370	1.758	0.8902
1.475 840	57.3	1.141	0.2155	1.151	2.007 720	1.751	0.4145	1.049	1.516 862	1.744	0.8906
1.440 216	56.0	1.132	0.2272	1.140	1.964 382	1.731	0.4381	1.030	1.488 829	1.729	0.8910
1.423 927	54.7	1.123	0.2394	1.130	1.922 953	1.710	0.4628	1.011	1.462 178	1.715	0.8915
1.399 891	53.4	1.113	0.2521	1.118	1.883 312	1.689	0.4888	0.9910	1.436 822	1.700	0.8919
1.377 032	52.2	1.103	0.2653	1.107	1.845 347	1.668	0.5160	0.9711	1.412 684	1.684	0.8923
1.355 082	50.9	1.093	0.2791	1.095	1.808 954	1.646	0.5446	0.9508	1.389 693	1.669	0.8928
1.334 577	49.7	1.083	0.2935	1.083	1.774 040	1.623	0.5747	0.9302	1.307 782	1.653	0.8932

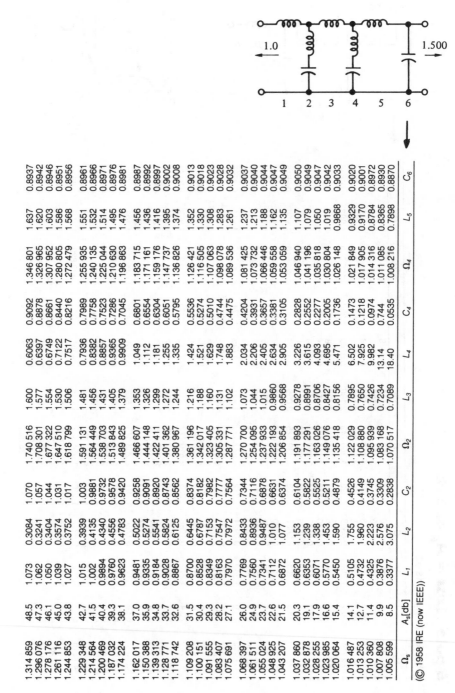

Ω_s	A_s[db]	L_1	L_2	C_2	Ω_2	L_3	L_4	C_4	Ω_4	L_5	C_6
1.314 859	48.5	1.073	0.3084	1.070	1.740 516	1.600	0.6063	0.9092	1.346 801	1.637	0.8937
1.296 076	47.3	1.062	0.3241	1.057	1.708 301	1.577	0.6397	0.8878	1.326 965	1.620	0.8942
1.278 176	46.1	1.050	0.3404	1.044	1.677 322	1.554	0.6749	0.8661	1.307 952	1.603	0.8946
1.261 116	45.0	1.039	0.3574	1.031	1.647 510	1.530	0.7122	0.8440	1.280 805	1.586	0.8951
1.244 853	43.8	1.027	0.3752	1.017	1.618 799	1.506	0.7517	0.8216	1.272 479	1.568	0.8956
1.229 348	42.7	1.015	0.3939	1.003	1.591 131	1.481	0.7936	0.7989	1.255 935	1.551	0.8961
1.214 564	41.5	1.002	0.4135	0.9881	1.564 449	1.456	0.8382	0.7758	1.240 135	1.532	0.8966
1.200 469	40.4	0.9894	0.4340	0.9732	1.538 703	1.431	0.8857	0.7523	1.225 044	1.514	0.8971
1.187 032	39.3	0.9760	0.4556	0.9578	1.513 843	1.405	0.9365	0.7286	1.210 630	1.495	0.8976
1.174 224	38.1	0.9623	0.4783	0.9420	1.489 825	1.379	0.9909	0.7045	1.196 863	1.476	0.8981
1.162 017	37.0	0.9481	0.5022	0.9258	1.466 607	1.353	1.049	0.6801	1.183 715	1.456	0.8987
1.150 388	35.9	0.9335	0.5274	0.9091	1.444 148	1.326	1.112	0.6554	1.171 161	1.436	0.8992
1.139 313	34.8	0.9184	0.5541	0.8920	1.422 411	1.299	1.181	0.6304	1.159 176	1.416	0.8997
1.128 771	33.7	0.9028	0.5824	0.8743	1.401 362	1.272	1.255	0.6051	1.147 737	1.395	0.9002
1.118 742	32.6	0.8867	0.6125	0.8562	1.380 967	1.244	1.335	0.5795	1.136 826	1.374	0.9008
1.109 208	31.5	0.8700	0.6445	0.8374	1.361 196	1.216	1.424	0.5536	1.126 421	1.352	0.9013
1.100 151	30.4	0.8528	0.6787	0.8182	1.342 017	1.188	1.521	0.5274	1.116 505	1.330	0.9018
1.091 555	29.3	0.8349	0.7153	0.7982	1.323 405	1.160	1.629	0.5010	1.107 063	1.308	0.9023
1.083 407	28.2	0.8163	0.7547	0.7777	1.305 331	1.131	1.748	0.4744	1.098 078	1.283	0.9028
1.075 691	27.1	0.7970	0.7972	0.7564	1.287 771	1.102	1.883	0.4475	1.089 536	1.261	0.9032
1.068 397	26.0	0.7769	0.8433	0.7344	1.270 700	1.073	2.034	0.4204	1.081 425	1.237	0.9037
1.061 511	24.9	0.7560	0.8936	0.7116	1.254 095	1.044	2.206	0.3931	1.073 732	1.213	0.9040
1.055 024	23.7	0.7341	0.9487	0.6878	1.237 933	1.015	2.405	0.3657	1.066 446	1.188	0.9044
1.048 925	22.6	0.7112	1.010	0.6631	1.222 193	0.9860	2.634	0.3381	1.059 558	1.162	0.9047
1.043 207	21.5	0.6872	1.077	0.6374	1.206 854	0.9568	2.905	0.3105	1.053 059	1.135	0.9049
1.037 860	20.3	0.6620	1.153	0.6104	1.191 893	0.9278	3.226	0.2828	1.046 940	1.107	0.9050
1.032 878	19.1	0.6353	1.239	0.5822	1.177 291	0.8991	3.615	0.2552	1.041 196	1.079	0.9049
1.028 255	17.9	0.6071	1.338	0.5525	1.163 026	0.8706	4.093	0.2277	1.035 818	1.050	0.9047
1.023 985	16.6	0.5770	1.453	0.5211	1.149 076	0.8427	4.695	0.2005	1.030 804	1.019	0.9042
1.020 064	15.4	0.5450	1.590	0.4879	1.135 418	0.8156	5.471	0.1736	1.026 148	0.9868	0.9033
1.016 487	14.1	0.5105	1.755	0.4526	1.122 029	0.7895	6.502	0.1473	1.021 849	0.9329	0.9020
1.013 253	12.7	0.4732	1.960	0.4149	1.108 880	0.7650	7.925	0.1218	1.017 905	0.9170	0.9001
1.010 360	11.4	0.4325	2.223	0.3745	1.095 939	0.7426	9.982	0.0974	1.014 316	0.8784	0.8972
1.007 808	9.9	0.3876	2.576	0.3309	1.083 168	0.7234	13.14	0.744	1.011 085	0.8365	0.8930
1.005 599	8.5	0.3377	3.075	0.2838	1.070 517	0.7089	18.40	0.0535	1.008 216	0.7998	0.8870

6 pole Loss = 0 dB at 0 Hz

Table A10.8 A_p = 0.18 dB

Ω_s	A_s[db]	C_1	C_2	L_2	Ω_2	C_3	C_4	L_4	Ω_4	C_5	L_6
3.878 298	112.5	1.138	0.0209	1.500	5.644 802	1.790	0.0350	1.769	4.020 935	1.500	1.158
3.655 090	109.3	1.135	0.0237	1.496	5.314 073	1.784	0.0396	1.761	3.788 961	1.496	1.158
3.456 975	108.3	1.132	0.0266	1.492	5.020 165	1.777	0.0445	1.751	3.583 033	1.492	1.158
3.279 996	103.4	1.129	0.0297	1.488	4.757 266	1.770	0.0497	1.742	3.399 040	1.488	1.158
3.120 982	100.7	1.125	0.0330	1.483	4.520 722	1.763	0.0552	1.731	3.233 693	1.483	1.158
2.977 369	98.1	1.122	0.0365	1.478	4.306 769	1.756	0.0611	1.720	3.084 330	1.479	1.158
2.847 060	95.6	1.118	0.0401	1.473	4.112 326	1.748	0.0673	1.709	2.948 774	1.474	1.157
2.728 322	93.3	1.114	0.0440	1.468	3.934 847	1.739	0.0738	1.697	2.825 225	1.469	1.157
2.619 709	91.0	1.110	0.0480	1.463	3.772 213	1.731	0.0807	1.685	2.712 184	1.464	1.157
2.520 009	88.8	1.106	0.0523	1.457	3.622 641	1.722	0.0879	1.672	2.608 393	1.458	1.157
2.428 196	86.7	1.102	0.0568	1.451	3.484 024	1.712	0.0955	1.658	2.512 785	1.452	1.157
2.343 395	84.6	1.097	0.0614	1.445	3.356 877	1.702	0.1035	1.644	2.424 454	1.446	1.156
2.264 858	82.6	1.092	0.0663	1.430	3.238 301	1.692	0.1118	1.630	2.342 621	1.440	1.156
2.191 939	80.7	1.087	0.0714	1.432	3.127 945	1.682	0.1205	1.615	2.266 617	1.433	1.156
2.124 078	78.9	1.082	0.0767	1.425	3.024 987	1.671	0.1297	1.599	2.195 860	1.427	1.156
2.080 787	77.1	1.077	0.0822	1.418	2.928 712	1.660	0.1392	1.583	2.129 845	1.420	1.155
2.001 642	75.3	1.071	0.0880	1.410	2.838 492	1.648	0.1492	1.567	2.068 129	1.413	1.155
1.946 266	73.6	1.065	0.0940	1.403	2.753 776	1.636	0.1597	1.550	2.010 323	1.403	1.155
1.894 331	72.0	1.059	0.1003	1.395	2.674 079	1.624	0.1706	1.532	1.956 085	1.398	1.154
1.845 543	70.4	1.053	0.1068	1.386	2.598 969	1.611	0.1820	1.514	1.905 110	1.390	1.154
1.799 643	68.8	1.047	0.1135	1.378	2.528 063	1.598	0.1939	1.496	1.857 129	1.382	1.154
1.756 398	67.3	1.040	0.1206	1.369	2.461 022	1.585	0.2063	1.477	1.811 902	1.374	1.153
1.715 603	65.8	1.033	0.1279	1.360	2.397 538	1.571	0.2192	1.457	1.769 212	1.365	1.153
1.677 070	64.3	1.026	0.1355	1.351	2.337 337	1.557	0.2328	1.437	1.728 868	1.356	1.152
1.640 634	62.8	1.019	0.1434	1.341	2.280 174	1.543	0.2469	1.417	1.690 696	1.348	1.152
1.606 142	61.4	1.012	0.1516	1.332	2.225 824	1.528	0.2617	1.396	1.654 538	1.338	1.151
1.573 460	60.0	1.004	0.1601	1.321	2.174 087	1.513	0.2772	1.374	1.620 254	1.329	1.151
1.542 462	58.7	0.9963	0.1689	1.311	2.124 779	1.498	0.2933	1.352	1.587 714	1.319	1.150
1.513 038	57.3	0.9882	0.1781	1.300	2.077 734	1.482	0.3103	1.330	1.556 804	1.309	1.150
1.485 086	56.0	0.9798	0.1877	1.289	2.032 800	1.466	0.3280	1.309	1.527 416	1.299	1.149
1.458 511	54.7	0.9712	0.1976	1.278	1.989 839	1.450	0.3465	1.284	1.498 453	1.289	1.148
1.433 230	53.4	0.9624	0.2079	1.266	1.948 725	1.433	0.3659	1.260	1.472 828	1.278	1.148
1.409 164	52.2	0.9533	0.2187	1.255	1.909 340	1.416	0.3863	1.235	1.447 459	1.267	1.147
1.386 241	50.9	0.9439	0.2298	1.242	1.871 578	1.399	0.4078	1.211	1.423 273	1.256	1.146
1.364 398	49.7	0.9343	0.2414	1.230	1.835 340	1.381	0.4303	1.185	1.400 200	1.245	1.146

Ω_s	L_1	A_s[db]	L_2	C_2	Ω_2	L_3	L_4	C_4	Ω_4	L_5	C_6
1.343 572	0.9244	48.5	0.2535	1.217	1.800 536	1.363	0.4540	1.160	1.378 179	1.234	1.145
1.323 710	0.9142	47.3	0.2661	1.204	1.767 082	1.345	0.4790	1.133	1.357 152	1.222	1.144
1.304 759	0.9037	46.1	0.2792	1.190	1.734 901	1.327	0.5054	1.107	1.337 064	1.210	1.143
1.286 672	0.8929	45.0	0.2929	1.176	1.703 919	1.308	0.5333	1.080	1.317 868	1.197	1.142
1.269 406	0.8819	43.8	0.3072	1.162	1.674 071	1.289	0.5628	1.052	1.299 518	1.185	1.141
1.252 921	0.8705	42.7	0.3221	1.147	1.645 294	1.269	0.5941	1.024	1.281 971	1.172	1.140
1.237 179	0.8587	41.5	0.3377	1.132	1.617 530	1.249	0.6274	0.9957	1.265 189	1.159	1.139
1.222 145	0.8466	40.4	0.3541	1.116	1.590 725	1.229	0.6629	0.9668	1.249 136	1.143	1.138
1.207 787	0.8342	39.3	0.3712	1.100	1.564 828	1.209	0.7008	0.9375	1.233 777	1.131	1.137
1.194 077	0.8214	38.1	0.3892	1.084	1.539 791	1.188	0.7413	0.9077	1.219 083	1.117	1.136
1.180 985	0.8081	37.0	0.4081	1.067	1.515 571	1.107	0.7848	0.8775	1.203 023	1.103	1.134
1.168 486	0.7945	35.9	0.4280	1.049	1.492 126	1.146	0.8317	0.8468	1.191 672	1.088	1.133
1.156 557	0.7804	34.8	0.4490	1.032	1.469 414	1.125	0.8823	0.8157	1.178 704	1.074	1.131
1.145 175	0.7659	33.7	0.4712	1.013	1.447 401	1.103	0.9372	0.7843	1.166 396	1.058	1.130
1.134 320	0.7509	32.6	0.4947	0.9940	1.426 049	1.081	0.9970	0.7324	1.154 626	1.043	1.128
1.123 973	0.7354	31.5	0.5196	0.9744	1.405 326	1.059	1.062	0.7201	1.143 375	1.026	1.126
1.114 116	0.7193	30.4	0.5462	0.9542	1.385 199	1.037	1.134	0.6874	1.132 624	1.010	1.125
1.104 733	0.7027	29.3	0.5746	0.9332	1.365 637	1.014	1.213	0.6543	1.122 356	0.9932	1.123
1.095 809	0.6854	28.2	0.6050	0.9115	1.346 613	0.9915	1.301	0.6208	1.112 555	0.9759	1.120
1.087 329	0.6674	27.1	0.6377	0.8891	1.328 096	0.9686	1.400	0.5870	1.103 207	0.9582	1.118
1.079 282	0.6488	26.0	0.6730	0.8657	1.310 060	0.9456	1.511	0.5528	1.094 297	0.9399	1.116
1.071 656	0.6293	24.9	0.7114	0.8415	1.292 478	0.9225	1.636	0.5184	1.085 815	0.9211	1.113
1.064 439	0.6089	23.7	0.7533	0.8162	1.273 324	0.8994	1.780	0.4836	1.077 747	0.9017	1.110
1.057 623	0.5876	22.6	0.7994	0.7898	1.258 571	0.8762	1.947	0.4486	1.070 085	0.8816	1.107
1.051 198	0.5652	21.5	0.8503	0.7621	1.242 193	0.8530	2.141	0.4134	1.062 820	0.8608	1.104
1.045 158	0.5417	20.3	0.9073	0.7331	1.226 164	0.8299	2.372	0.3781	1.055 943	0.8393	1.100
1.039 495	0.5168	19.1	0.9716	0.7025	1.210 456	0.8071	2.650	0.3426	1.049 447	0.8168	1.096
1.034 204	0.4905	17.9	1.045	0.6701	1.195 041	0.7845	2.990	0.3072	1.043 327	0.7932	1.091
1.029 281	0.4624	16.6	1.130	0.6358	1.179 887	0.7625	3.415	0.2720	1.037 578	0.7685	1.086
1.024 722	0.4323	15.4	1.230	0.5991	1.164 960	0.7411	3.961	0.2370	1.032 198	0.7423	1.080
1.020 525	0.3999	14.1	1.350	0.5598	1.150 224	0.7206	4.677	0.2026	1.027 183	0.7144	1.073
1.016 691	0.3648	12.7	1.499	0.5174	1.135 632	0.7016	5.659	0.1690	1.022 536	0.6843	1.064
1.013 219	0.3263	11.4	1.687	0.4715	1.121 129	0.6845	7.062	0.1366	1.018 256	0.6518	1.055
1.010 114	0.2837	9.9	1.938	0.4214	1.106 645	0.6702	9.190	0.1058	1.014 351	0.6158	1.043
1.007 381	0.2358	8.5	2.288	0.3664	1.092 084	0.6603	12.67	0.0772	1.010 827	0.5750	1.027

7 pole
Table A10.9 $A_p = 0.18$ dB

Ω_s	A_s[db]	C_1	C_2	L_2	Ω_2	C_3	C_4	L_4	Ω_4	C_5	C_6	L_6	Ω_6	C_7
2.281 172	105.4	1.310	0.0290	1.358	5.038 750	2.100	0.1353	1.357	2.333 900	2.049	0.0955	1.281	2.850 592	1.247
2.202 689	103.0	1.308	0.0314	1.355	4.848 897	2.089	0.1465	1.345	2.253 156	2.034	0.1034	1.272	2.756 829	1.240
2.130 054	100.7	1.306	0.0339	1.353	4.672 457	2.078	0.1582	1.332	2.178 400	2.019	0.1117	1.263	2.661 529	1.233
2.062 665	98.5	1.304	0.0364	1.350	4.508 037	2.066	0.1704	1.319	2.109 040	2.003	0.1204	1.254	2.572 921	1.226
2.000 000	96.3	1.302	0.0391	1.347	4.354 434	2.054	0.1833	1.305	2.044 515	1.987	0.1295	1.245	2.490 337	1.218
1.941 604	94.2	1.299	0.0420	1.344	4.210 595	2.042	0.1966	1.292	1.984 368	1.970	0.1390	1.235	2.413 194	1.210
1.887 080	92.2	1.297	0.0449	1.341	4.075 602	2.029	0.2106	1.277	1.928 190	1.952	0.1490	1.225	2.340 984	1.202
1.836 078	90.2	1.294	0.0479	1.338	3.948 647	2.016	0.2252	1.262	1.875 623	1.934	0.1593	1.214	2.273 259	1.193
1.788 292	88.3	1.292	0.0511	1.335	3.829 016	2.002	0.2404	1.247	1.826 351	1.916	0.1702	1.204	2.209 625	1.184
1.743 447	86.4	1.280	0.0544	1.332	3.716 076	1.988	0.2562	1.232	1.780 095	1.807	0.1815	1.193	2.149 731	1.175
1.701 302	84.6	1.286	0.0578	1.328	3.609 267	1.973	0.2727	1.216	1.736 606	1.878	0.1932	1.181	2.093 268	1.165
1.661 640	82.8	1.283	0.0614	1.324	3.508 087	1.959	0.2900	1.199	1.695 662	1.858	0.2055	1.169	2.039 957	1.155
1.624 269	81.0	1.280	0.0650	1.321	3.412 086	1.943	0.3079	1.183	1.657 065	1.837	0.2183	1.157	1.989 552	1.145
1.589 016	79.3	1.277	0.0689	1.317	3.320 862	1.928	0.3267	1.165	1.620 638	1.817	0.2317	1.145	1.941 830	1.135
1.555 724	77.6	1.274	0.0728	1.313	3.234 050	1.912	0.3462	1.148	1.586 220	1.795	0.2456	1.132	1.896 591	1.124
1.524 253	76.0	1.270	0.0770	1.308	3.151 325	1.895	0.3666	1.130	1.553 668	1.773	0.2601	1.119	1.853 653	1.113
1.494 477	74.3	1.267	0.0812	1.304	3.072 388	1.879	0.3879	1.112	1.522 851	1.751	0.2753	1.105	1.812 855	1.102
1.466 279	72.8	1.263	0.0857	1.300	2.996 969	1.862	0.4104	1.093	1.493 651	1.728	0.2911	1.092	1.774 048	1.090
1.439 557	71.2	1.259	0.0903	1.295	2.924 824	1.844	0.4332	1.074	1.465 961	1.705	0.3076	1.077	1.737 098	1.078
1.414 214	69.7	1.255	0.0950	1.290	2.855 727	1.826	0.4575	1.055	1.439 683	1.682	0.3248	1.063	1.701 881	1.066
1.390 164	68.2	1.251	0.1000	1.285	2.789 476	1.808	0.4828	1.035	1.414 728	1.657	0.3428	1.048	1.668 286	1.053
1.367 327	66.7	1.247	0.1051	1.280	2.725 881	1.789	0.5093	1.015	1.391 016	1.633	0.3617	1.033	1.636 211	1.040
1.345 633	65.2	1.243	0.1105	1.275	2.664 770	1.770	0.5370	0.9944	1.368 471	1.608	0.3814	1.017	1.605 563	1.027
1.325 013	63.7	1.238	0.1160	1.269	2.605 984	1.751	0.5661	0.9736	1.347 026	1.583	0.4020	1.001	1.576 255	1.013
1.305 407	62.3	1.234	0.1217	1.264	2.549 377	1.731	0.5965	0.9525	1.326 618	1.557	0.4235	0.9850	1.548 208	0.9992
1.286 760	60.9	1.229	0.1277	1.258	2.494 813	1.711	0.6286	0.9310	1.307 190	1.531	0.4462	0.9684	1.521 349	0.9848
1.269 018	59.5	1.224	0.1339	1.252	2.442 167	1.690	0.6622	0.9093	1.288 687	1.504	0.4699	0.9514	1.495 612	0.9699
1.252 136	58.1	1.219	0.1404	1.246	2.391 323	1.669	0.6977	0.8872	1.271 063	1.477	0.4948	0.9340	1.470 934	0.9547
1.236 068	56.8	1.213	0.1471	1.239	2.342 170	1.648	0.7351	0.8648	1.254 270	1.450	0.5211	0.9163	1.447 259	0.9391
1.220 775	55.4	1.208	0.1541	1.232	2.294 610	1.626	0.7745	0.8420	1.238 269	1.422	0.5487	0.8981	1.424 533	0.9230
1.206 218	54.1	1.202	0.1614	1.225	2.248 546	1.604	0.8163	0.8190	1.223 020	1.394	0.5778	0.8796	1.402 707	0.9065
1.192 363	52.7	1.196	0.1690	1.218	2.203 891	1.581	0.8605	0.7957	1.208 487	1.365	0.6085	0.8607	1.381 735	0.8896
1.179 178	51.4	1.190	0.1770	1.211	2.160 560	1.558	0.9075	0.7721	1.194 638	1.336	0.6411	0.8414	1.361 575	0.8722
1.166 633	50.1	1.183	0.1853	1.203	2.118 476	1.535	0.9576	0.7482	1.181 442	1.307	0.6755	0.8217	1.342 188	0.8543
1.154 701	48.8	1.177	0.1939	1.195	2.077 565	1.511	1.011	0.7240	1.168 869	1.278	0.7121	0.8016	1.323 537	0.8360

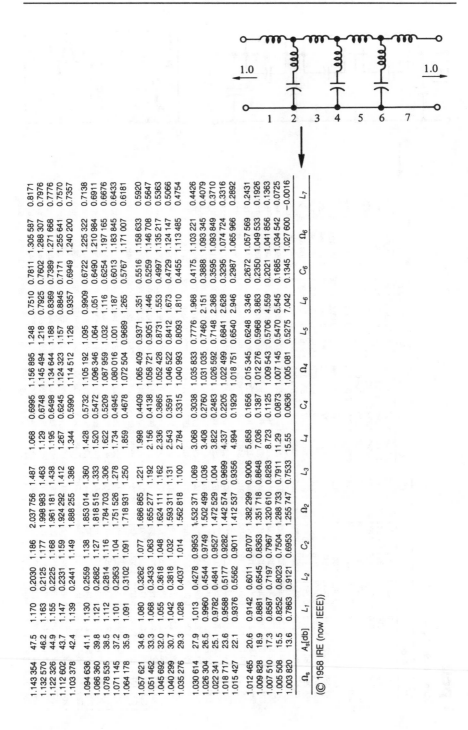

Ω_s	A_s(db)	L_1	L_2	C_2	Ω_2	L_3	L_4	C_4	Ω_4	L_5	L_6	C_6	Ω_6	L_7
1.143 354	47.5	1.170	0.2030	1.186	2.037 756	1.487	1.068	0.6995	1.156 895	1.248	0.7510	0.7811	1.305 587	0.8171
1.132 570	46.2	1.163	0.2125	1.177	1.998 983	1.463	1.129	0.6748	1.145 494	1.218	0.7925	0.7602	1.288 307	0.7976
1.122 326	44.9	1.155	0.2225	1.168	1.961 181	1.438	1.195	0.6498	1.134 644	1.188	0.8369	0.7389	1.271 668	0.7776
1.112 602	43.7	1.147	0.2331	1.159	1.924 292	1.412	1.267	0.6245	1.124 323	1.157	0.8845	0.7171	1.255 641	0.7570
1.103 378	42.4	1.139	0.2441	1.149	1.888 255	1.386	1.344	0.5990	1.114 512	1.126	0.9357	0.6949	1.240 200	0.7357
1.094 636	41.1	1.130	0.2559	1.138	1.853 014	1.360	1.428	0.5732	1.105 192	1.095	0.9909	0.6722	1.225 322	0.7138
1.086 360	39.8	1.121	0.2682	1.127	1.818 515	1.333	1.520	0.5472	1.096 346	1.064	1.051	0.6490	1.210 984	0.6911
1.078 535	38.5	1.112	0.2814	1.116	1.784 703	1.306	1.622	0.5209	1.087 959	1.032	1.116	0.6254	1.197 165	0.6676
1.071 145	37.2	1.101	0.2953	1.104	1.751 526	1.278	1.734	0.4945	1.080 016	1.001	1.187	0.6013	1.183 845	0.6433
1.064 178	35.9	1.091	0.3102	1.091	1.718 931	1.250	1.859	0.4678	1.072 504	0.9689	1.265	0.5767	1.171 007	0.6181
1.057 621	34.6	1.080	0.3262	1.077	1.686 865	1.221	1.998	0.4409	1.065 409	0.9371	1.351	0.5516	1.158 633	0.5920
1.051 462	33.3	1.068	0.3433	1.063	1.655 277	1.192	2.156	0.4138	1.058 721	0.9051	1.446	0.5259	1.146 708	0.5647
1.045 692	32.0	1.055	0.3618	1.048	1.624 111	1.162	2.336	0.3865	1.052 428	0.8731	1.553	0.4997	1.135 217	0.5363
1.040 299	30.7	1.042	0.3818	1.032	1.593 311	1.131	2.543	0.3591	1.046 522	0.8412	1.673	0.4729	1.124 147	0.5066
1.035 276	29.3	1.028	0.4037	1.014	1.562 818	1.100	2.784	0.3315	1.040 993	0.8093	1.810	0.4455	1.113 485	0.4754
1.030 614	27.9	1.013	0.4278	0.9953	1.532 371	1.069	3.068	0.3038	1.035 833	0.7776	1.968	0.4175	1.103 221	0.4426
1.026 304	26.5	0.9960	0.4544	0.9749	1.502 499	1.036	3.408	0.2760	1.031 035	0.7460	2.151	0.3888	1.093 345	0.4079
1.022 341	25.1	0.9782	0.4841	0.9527	1.472 529	1.004	3.822	0.2483	1.026 592	0.7148	2.368	0.3595	1.083 849	0.3710
1.018 717	23.6	0.9588	0.5177	0.9282	1.442 574	0.9699	4.337	0.2205	1.022 499	0.6841	2.628	0.3295	1.074 724	0.3316
1.015 427	22.1	0.9376	0.5562	0.9011	1.412 537	0.9356	4.994	0.1929	1.018 751	0.6540	2.946	0.2987	1.065 966	0.2892
1.012 465	20.6	0.9142	0.6011	0.8707	1.382 299	0.9006	5.858	0.1656	1.015 345	0.6248	3.346	0.2672	1.057 569	0.2431
1.009 828	18.9	0.8881	0.6545	0.8363	1.351 718	0.8648	7.036	0.1387	1.012 276	0.5968	3.863	0.2350	1.049 533	0.1926
1.007 510	17.3	0.8587	0.7197	0.7967	1.320 610	0.8283	8.723	0.1125	1.009 543	0.5706	4.559	0.2021	1.041 856	0.1363
1.005 508	15.5	0.8252	0.8023	0.7504	1.288 733	0.7911	11.29	0.0873	1.007 145	0.5470	5.545	0.1685	1.034 542	0.0725
1.003 820	13.6	0.7863	0.9121	0.6953	1.255 747	0.7533	15.55	0.0636	1.005 081	0.5275	7.042	0.1345	1.027 600	−0.0016

9 pole

Table A10.10 $A_p = 0.18$ dB

Ω_s	A_s[db]	C_1	C_2	L_2	Ω_2	C_3	C_4	L_4	Ω_4	C_5	C_6
1.701 302	116.1	1.318	0.0334	1.367	4.543 863	2.067	0.2078	1.310	1.916 432	1.934	0.2703
1.661 640	113.8	1.316	0.0376	1.365	4.414 407	2.055	0.2207	1.297	1.869 139	1.912	0.2873
1.624 269	111.5	1.315	0.0399	1.362	2.291 507	2.043	0.2341	1.283	1.824 497	1.889	0.3050
1.580 016	109.3	1.313	0.0422	1.360	4.174 652	2.030	0.2481	1.269	1.782 266	1.866	0.3234
1.555 724	107.2	1.310	0.0446	1.357	4.063 382	2.017	0.2626	1.254	1.742 285	1.842	0.3426
1.524 253	105.1	1.308	0.0471	1.355	3.957 281	2.004	0.2777	1.240	1.704 392	1.817	0.3626
1.494 477	103.0	1.306	0.0498	1.352	3.855 969	1.991	0.2934	1.224	1.668 439	1.792	0.3834
1.466 279	100.9	1.304	0.0525	1.349	3.759 105	1.977	0.3097	1.209	1.634 294	1.767	0.4052
1.439 557	98.9	1.301	0.0553	1.346	3.666 376	1.963	0.3267	1.193	1.601 835	1.741	0.4278
1.414 214	97.0	1.299	0.0582	1.343	3.577 497	1.948	0.3444	1.177	1.570 952	1.714	0.4515
1.390 164	93.0	1.296	0.0612	1.340	3.492 207	1.934	0.3628	1.160	1.541 544	1.687	0.4762
1.367 327	93.1	1.294	0.0643	1.336	3.410 268	1.98	0.3820	1.143	1.513 520	1.659	0.5020
1.345 633	91.2	1.291	0.0676	1.333	3.331 459	1.903	0.4019	1.126	1.486 796	1.631	0.5289
1.325 013	89.3	1.288	0.0710	1.329	3.255 578	1.887	0.4227	1.108	1.461 293	1.603	0.5571
1.305 407	87.5	1.285	0.0745	1.326	3.182 438	1.871	0.4444	1.090	1.436 942	1.574	0.5867
1.286 760	85.7	1.282	0.0781	1.322	3.111 863	1.854	0.4671	1.071	1.413 677	1.544	0.6176
1.269 018	83.9	1.279	0.0810	1.318	3.043 699	1.837	0.4908	1.032	1.391 438	1.514	0.6501
1.252 136	82.1	1.275	0.0858	1.314	2.977 790	1.820	0.5155	1.033	1.370 170	1.484	0.6843
1.236 068	80.4	1.272	0.0899	1.310	2.914 000	1.802	0.5414	1.014	1.349 821	1.453	0.7202
1.220 775	78.6	1.268	0.0942	1.305	2.852 198	1.784	0.5685	0.0939	1.330 344	1.421	0.7580
1.206 218	76.9	1.265	0.0986	.301	2.792 263	1.765	0.5969	0.9737	1.311 695	1.389	0.7979
1.102 363	75.2	1.261	0.1032	1.296	2.734 079	1.746	0.6268	0.9531	1.293 834	1.357	0.8401
1.179 178	73.5	1.257	0.1080	1.291	2.677 540	1.726	0.6582	0.9321	1.276 723	1.324	0.8847
1.166 633	71.8	1.253	0.1131	1.286	2.622 544	1.707	0.6912	0.9108	1.260 327	1.291	0.9321
1.154 701	70.1	1.248	0.1183	1.281	2.568 993	1.686	0.7261	0.8891	1.244 613	1.257	0.9825
1.143 354	68.5	1.244	0.1238	1.275	2.516 797	1.666	0.7629	0.8670	1.229 551	1.223	1.036
1.132 570	66.8	1.239	0.1296	1.269	2.265 867	1.644	0.8019	0.8446	1.215 114	1.189	1.093
1.122 326	65.2	1.234	0.1356	1.263	2.416 121	1.623	0.8433	0.8217	1.201 275	1.154	1.155
1.112 602	63.5	1.229	0.1420	1.257	2.367 476	1.600	0.8873	0.7985	1.188 009	1.119	1.221
1.103 378	61.9	1.223	0.1487	1.250	2.319 854	1.578	0.9342	0.7749	1.175 295	1.083	1.292
1.094 636	60.2	1.217	0.1557	1.243	2.273 180	1.554	0.9844	0.7509	1.163 112	1.047	1.369
1.086 360	58.6	1.211	0.1631	1.236	2.227 378	1.531	1.038	0.7265	1.151 440	1.011	1.453
1.078 535	56.9	1.205	0.1710	1.228	2.182 375	1.506	1.096	0.7017	1.140 260	0.9738	1.544
1.071 145	55.2	1.198	0.1703	1.220	2.138 097	1.481	1.159	0.6764	1.129 558	0.9367	1.644
1.064 178	53.6	1.101	0.1882	1.211	2.094 470	1.455	1.227	0.6507	1.119 316	0.8992	1.754
1.057 621	51.9	1.184	0.1977	1.202	2.051 420	1.429	1.301	0.6245	1.109 521	0.8614	1.876
1.051 462	50.2	1.176	0.2078	1.192	2.008 869	1.401	1.382	0.5979	1.100 160	0.8233	2.013
1.045 692	48.5	1.167	0.2187	1.182	1.966 738	1.373	1.471	0.5708	1.091 222	0.7849	2.166
1.040 209	46.8	1.158	0.2305	1.171	1.924 942	1.344	1.571	0.5432	1.082 095	0.7463	2.339
1.035 276	40.1	1.148	0.2433	1.159	1.883 393	1.314	1.082	0.5160	1.074 570	0.7073	2.538
1.030 614	43.3	1.137	0.2572	1.140	1.841 992	1.283	1.807	0.4862	1.066 839	0.6681	2.768
1.026 304	41.5	1.126	0.2724	1.132	1.800 631	1.251	1.950	0.4569	1.059 494	0.6287	3.036
1.022 341	39.6	1.113	0.2803	1.117	1.750 188	1.218	2.115	0.4268	1.052 530	0.5891	3.355
1.018 717	37.7	1.099	0.3081	1.100	1.717 524	1.183	2.308	0.3961	1.045 943	0.5493	3.741
1.015 427	35.8	1.084	0.3202	1.082	1.675 471	1.140	2.538	0.3645	1.039 728	0.5094	4.216
1.012 465	33.8	1.067	0.3534	1.061	1.632 828	1.108	2.817	0.3321	1.033 885	0.4693	4.817
1.009 828	31.7	1.047	0.3814	1.038	1.589 344	1.067	3.166	0.2986	1.028 414	0.4293	5.599
1.007 510	29.5	1.025	0.4145	1.011	1.544 692	1.024	3.616	0.2641	1.023 319	0.3894	6.660
1.005 508	27.1	0.9995	0.4548	0.9794	1.498 431	0.9782	4.223	0.2282	1.018 605	0.3496	8.173
1.003 820	24.6	0.9688	0.5054	0.9411	1.449 932	0.9284	5.093	0.1909	1.014 284	0.3103	10.50
Ω_s	A_s[db]	L_1	L_2	C_2	Ω_2	L_3	L_4	C_4	Ω_4	L_5	L_6

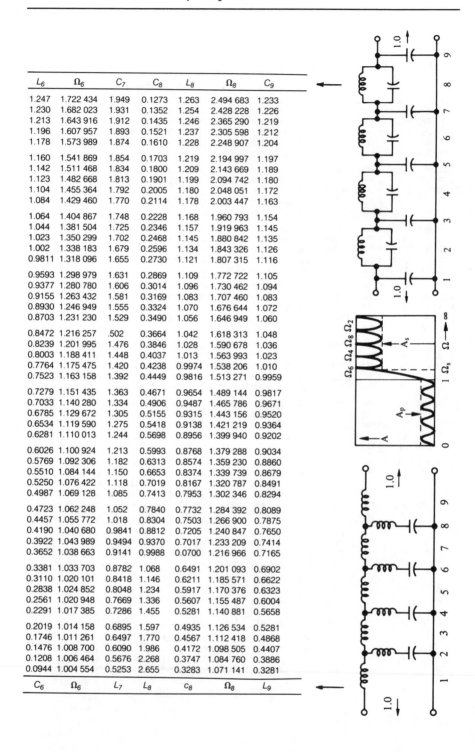

L_6	Ω_6	C_7	C_8	L_8	Ω_8	C_9
1.247	1.722 434	1.949	0.1273	1.263	2.494 683	1.233
1.230	1.682 023	1.931	0.1352	1.254	2.428 228	1.226
1.213	1.643 916	1.912	0.1435	1.246	2.365 290	1.219
1.196	1.607 957	1.893	0.1521	1.237	2.305 598	1.212
1.178	1.573 989	1.874	0.1610	1.228	2.248 907	1.204
1.160	1.541 869	1.854	0.1703	1.219	2.194 997	1.197
1.142	1.511 468	1.834	0.1800	1.209	2.143 669	1.189
1.123	1.482 668	1.813	0.1901	1.199	2.094 742	1.180
1.104	1.455 364	1.792	0.2005	1.180	2.048 051	1.172
1.084	1.429 460	1.770	0.2114	1.178	2.003 447	1.163
1.064	1.404 867	1.748	0.2228	1.168	1.960 793	1.154
1.044	1.381 504	1.725	0.2346	1.157	1.919 963	1.145
1.023	1.350 299	1.702	0.2468	1.145	1.880 842	1.135
1.002	1.338 183	1.679	0.2596	1.134	1.843 326	1.126
0.9811	1.318 096	1.655	0.2730	1.121	1.807 315	1.116
0.9593	1.298 979	1.631	0.2869	1.109	1.772 722	1.105
0.9377	1.280 780	1.606	0.3014	1.096	1.730 462	1.094
0.9155	1.263 432	1.581	0.3169	1.083	1.707 460	1.083
0.8930	1.246 949	1.555	0.3324	1.070	1.676 644	1.072
0.8703	1.231 230	1.529	0.3490	1.056	1.646 949	1.060
0.8472	1.216 257	.502	0.3664	1.042	1.618 313	1.048
0.8239	1.201 995	1.476	0.3846	1.028	1.590 678	1.036
0.8003	1.188 411	1.448	0.4037	1.013	1.563 993	1.023
0.7764	1.175 475	1.420	0.4238	0.9974	1.538 206	1.010
0.7523	1.163 158	1.392	0.4449	0.9816	1.513 271	0.9959
0.7279	1.151 435	1.363	0.4671	0.9654	1.489 144	0.9817
0.7033	1.140 280	1.334	0.4906	0.9487	1.465 786	0.9671
0.6785	1.129 672	1.305	0.5155	0.9315	1.443 156	0.9520
0.6534	1.119 590	1.275	0.5418	0.9138	1.421 219	0.9364
0.6281	1.110 013	1.244	0.5698	0.8956	1.399 940	0.9202
0.6026	1.100 924	1.213	0.5993	0.8768	1.379 288	0.9034
0.5769	1.092 306	1.182	0.6313	0.8574	1.359 230	0.8860
0.5510	1.084 144	1.150	0.6653	0.8374	1.339 739	0.8679
0.5250	1.076 422	1.118	0.7019	0.8167	1.320 787	0.8491
0.4987	1.069 128	1.085	0.7413	0.7953	1.302 346	0.8294
0.4723	1.062 248	1.052	0.7840	0.7732	1.284 392	0.8089
0.4457	1.055 772	1.018	0.8304	0.7503	1.266 900	0.7875
0.4190	1.040 680	0.9841	0.8812	0.7205	1.240 847	0.7650
0.3922	1.043 989	0.9494	0.9370	0.7017	1.233 209	0.7414
0.3652	1.038 663	0.9141	0.9988	0.0700	1.216 966	0.7165
0.3381	1.033 703	0.8782	1.068	0.6491	1.201 093	0.6902
0.3110	1.020 101	0.8418	1.146	0.6211	1.185 571	0.6622
0.2838	1.024 852	0.8048	1.234	0.5917	1.170 376	0.6323
0.2561	1.020 948	0.7669	1.336	0.5607	1.155 487	0.6004
0.2291	1.017 385	0.7286	1.455	0.5281	1.140 881	0.5658
0.2019	1.014 158	0.6895	1.597	0.4935	1.126 534	0.5281
0.1746	1.011 261	0.6497	1.770	0.4567	1.112 418	0.4868
0.1476	1.008 700	0.6090	1.986	0.4172	1.098 505	0.4407
0.1208	1.006 464	0.5676	2.268	0.3747	1.084 760	0.3886
0.0944	1.004 554	0.5253	2.655	0.3283	1.071 141	0.3281
C_6	Ω_6	L_7	L_8	c_8	Ω_8	L_9

Appendix 11
Screening

The following information on screening is reproduced by courtesy of RFI Shielding Ltd, from their *Materials Design Manual*.

THE NEED TO CONTROL EMC

The result of failure to achieve EMC can range from mild annoyance through serious disruption of legitimate activities, to health and safety hazards. The security of information being processed by electronic means is a vital commercial and military interest, often referred to by the word TEMPEST. The electrical signals corresponding to the information may leak, by radiation or conduction, from the processing equipment and be intercepted by suitable sensitive receiving equipment.

Automotive electronics has extended, for example, into engine management and anti-skid braking systems. There are evident safety implications if such electronic devices malfunction when the vehicle is subject to legitimate RF fields from on-board or nearby radio transmitters.

Finally, the explosion of nuclear devices results in an intense burst of radio energy in the HF band which, at distances beyond the likelihood of thermal blast damage, can cause temporary malfunction or permanent damage to electronic equipment. This is known as EMP, Electro-Magnetic Pulse.

Most countries recognise the need to control EMC and have civil EMC specifications which must be met internally and also by importers of electronic equipment. These specifications mainly control the level of emissions from the equipment but it will not be long before the susceptibility of civil equipments to externally produced electro-magnetic energy will be controlled by specification. Military EMC specifications have long covered both emissions and susceptibility.

SOURCES OF EMC PROBLEMS AND THEIR CONTAINMENT

The operation of all electrical or electronic devices involves the changing of voltage or current levels intermittently or continuously, sometimes at fast rates. This results in the development of electro-magnetic energy at discrete frequencies and over bands of frequencies. In general, the circuit will radiate this energy into the space around it and also conduct the energy into the wiring, perhaps to emerge along the external power, signal or control lines.

Figure 2 shows the role of enclosure screening in limiting the coupling of unwanted radiation to a victim equipment. Figure 2 also shows how that victim equipment can be protected against external RF fields.

These notes do not go into any detail of the limitation of conducted interference by cable screening and filtering. The information is readily available from specialist manufacturers of line filters, screened cable and screened and filtered connectors. However, the attention is drawn, at the appropriate point, to the need for integrating all EMC measures to ensure the required results. For example, the necessary penetration of a screening enclosure by a screened cable or the output from an electrical filter, requires meticulous attention to achieving a low impedance electrical bond between enclosure and connector body or filter body.

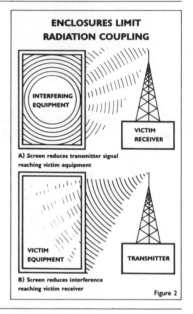

ENCLOSURES LIMIT RADIATION COUPLING

A) Screen reduces transmitter signal reaching victim equipment

B) Screen reduces interference reaching victim receiver

Figure 2

E-M WAVE HITS METALLIC BARRIER

Figure 3 shows, in general terms, what happens when an electro-magnetic wave strikes a metallic barrier.

The incoming wave has two components, an electric field and a magnetic field, at right-angles to each other and the direction of travel. The relative strengths of the two fields will be detailed later. Consider just the electric field which has strength E_i when it hits the barrier. Some of the energy is reflected back, strength E_r, but some carries on into the barrier, initially at strength $E_i t$.

This transmitted component gets absorbed as it travels through the barrier and arrives at the second face at strength $E_2 i$. Once more the energy divides into a reflected component $E_2 r$ and a transmitted component $E0$. The 'E' field screening effectiveness is defined as . . .

The use of the decibel is convenient to cope with the wide range of values encountered. A very modest screen might reduce the emergent field to one-tenth of the incident value, ie a screening effectiveness of 20dB. On the other hand a demanding application might require a reduction to one hundred thousandth of the incident field – a screening effectiveness of 100dB.

The incident 'H' field also suffers reflection and absorption as it passes through the front and back faces of the barrier, just like the 'E' field. However, the relative amounts are usually different as will be seen.

It is convenient to define screening effectiveness as

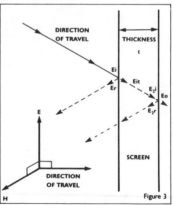

Figure 3

the sum of three terms, each expressed in dB, and have a closer look at the actual values of these terms.

S = SCREENING EFFECTIVENESS (dB)	dB	Percentage reduction
A = Absorption loss (dB)	0	0
R = Reflection loss (dB)	20	90
B = Correction factor (dB) (for multiple reflections	40	99
in thin screens)	60	99.9
S = A + R + B	80	99.99
	100	99.999, etc

Figure 4

ABSORPTION LOSS Figure 5 shows that absorption loss depends on the thickness of the barrier, the frequency of interest and two properties of the barrier material that is, the conductivity and the permeability, relative to copper. The table shows values for typical materials of interest.

Figure 5

$$A = 0.1315.t \sqrt{f.\sigma.\mu} (dB)$$

t = screen thickness (mm)
f = frequency (Hz)
σ = conductivity relative to copper
μ = permeability relative to copper

Note: For screen thickness (t) in inches replace the constant 0.1315 with 3.34

Material	σ	μ
Copper	1.00	1
Aluminium	0.61	1
Brass	0.61	1
Tin	0.15	1
Steel (SAE 1045)	0.10	1000
Monel	0.04	1
Stainless steel	0.02	500
Electroless nickel	0.02	1

Figure 6 shows the variation of absorption loss with frequency for two typical screening materials, copper and steel. Two thicknesses are considered 5mm (0.200″) and 0.5mm (0.020″).

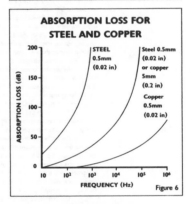

ABSORPTION LOSS FOR STEEL AND COPPER

Figure 6

REFLECTION LOSS (PLANE WAVE) The reflection loss increases with the ratio of the impedance of the incident wave to the impedance of the screen material. For plane EM waves, such as exist beyond a distance of about one-sixth of a wavelength from the source, the wave impedance is constant at about 377 ohms. The impedance of the screen material is proportional to the square root of the frequency times the permeability divided by the conductivity. Good conductors and non-magnetic materials give low screen impedance and hence high reflection loss. Working at higher frequencies raises the screen impedance and lowers the reflection loss. Figure 7 shows some typical values for reflection loss.

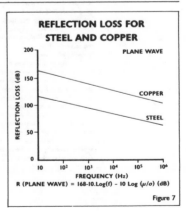

REFLECTION LOSS FOR STEEL AND COPPER

R (PLANE WAVE) = 168-10.Log(f) – 10 Log (μ/σ) (dB)

Figure 7

COMBINED ABSORPTION AND REFLECTION LOSS FOR PLANE WAVES

Figure 8 shows the total shielding effectiveness for a copper screen 0.5mm (0.02″) thick, in the far field, where the wave front is plane and the wave impedance is constant at 377 ohms. The poor absorption at low frequencies is compensated by the high reflection loss. The multiple reflection correction factor, B, is normally neglected for electric fields because the reflection loss is so large. This point will be considered later.

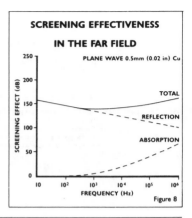

SCREENING EFFECTIVENESS IN THE FAR FIELD

PLANE WAVE 0.5mm (0.02 in) Cu

Figure 8

REFLECTION LOSS IN THE NEAR FIELD

The wave impedance in the near field depends on the nature of the source of the wave and the distance from that source. Figure 9 shows that for a rod or straight wire antenna, the wave impedance is high near the source. The impedance falls with distance from the source and levels out at the plane wave impedance value of 377 ohms. In contrast, if the source is a small wire loop, the field is predominantly magnetic and the wave impedance is low near the source. The impedance rises with distance away from the source but will also level at the free space value at distances beyond about one-sixth wavelength.

As detailed in the 'Enclosure Design' section, EMI shields are required in a range of materials for reasons other than those of attenuation alone. Such factors as compatibility with existing materials, physical strength and corrosion resistance, are all relevant. The properties of those materials used by RFI Shielding Ltd., are discussed here to assist in selection of the most suitable with regard to these factors. Comparative tables are provided at the end of the section.

Remembering that reflection loss varies as the ratio of wave to screen impedance it can be seen that reflection loss will depend on the type of wave being dealt with and how far the screen is from the source. For small, screened, equipments we are usually working in the near field and have to deal with this more complex situation. Figure 10 shows the relevant formulae.

The procedure for calculating the correction factor, B, is also shown in Figure 10. This is normally only calculated for the near-field magnetic case and then only if the absorption loss is less than 10dB. Re-reflection within the barrier, in the absence of much absorption, results in more energy passing through the second face of the barrier. Thus the correction factor is negative indicating a reduced screening effectiveness.

Figure 11 illustrates the variation of reflection loss with distance and frequency in the near field for a copper screen. Notice that in the near-field, as reflection loss for electric fields is higher, the closer the screen is to the source. For magnetic fields the reverse is true.

The electronic design engineer can therefore specify his screening requirements in terms of the emission frequency range of interference sources, their location relative to the screening effectiveness to be achieved.

The mechanical design engineer can then begin to explore screening enclosure material options and calculate their screening effectiveness.

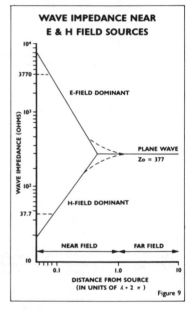

WAVE IMPEDANCE NEAR E & H FIELD SOURCES

E-FIELD DOMINANT

PLANE WAVE
$Z_o = 377$

H-FIELD DOMINANT

NEAR FIELD   FAR FIELD

DISTANCE FROM SOURCE
(IN UNITS OF $\lambda \div 2\pi$)

Figure 9

REFLECTION LOSS IN THE NEAR FIELD

R (Electric) = $321.8 - 20.\text{Log}(r)$
　　　　　　$- 30.\text{Log}(f) - 10.\text{Log}(\mu/\sigma)$ (dB)
R (Magnetic) = $14.6 + 20.\text{Log}(r)$
　　　　　　$+ 10.\text{Log}(f) + 10.\text{Log}(\sigma/\mu)$ (dB)

r	=	distance from source to screen (m)
f	=	frequency (Hz)
μ	=	permeability relative to copper
σ	=	conductivity relative to copper

CORRECTION FACTOR B

B	=	$20.\text{Log}(1 - \exp(-2.t/\delta))$ (dB)
t	=	screen thickness (mm)
δ	=	skin depth
	=	$0.102 \div \sqrt{f.\mu.\sigma}$ (mm)
For (t/δ)	=	0.1, B = -15dB
	=	0.5, = -4dB
	=	1.0, = -1dB

Figure 10

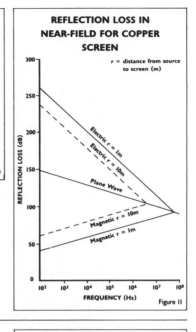

REFLECTION LOSS IN NEAR-FIELD FOR COPPER SCREEN

r = distance from source to screen (m)

Electric r = 1m
Electric r = 10m
Plane Wave
Magnetic r = 10m
Magnetic r = 1m

REFLECTION LOSS (dB)

FREQUENCY (Hz)

Figure 11

SCREEN MATERIALS

The provision of high screening effectiveness at very low frequencies can only be achieved by high permeability materials. The permeability of these materials falls off with frequency and can also be reduced if the incident magnetic field is high. Further, the permeability may be reduced by the mechanical working of the metal necessary to fabricate the required shape of screen. For all these reasons the exploitation of high permeability materials for screening purposes is a demanding task and recourse should be made to a specialist supplier in this field.

On the other hand, at higher frequencies it becomes possible to use cheaper metallic materials at quite modest thickness. Some typical screen materials are listed in Figure 12. Depending on the screening effectiveness requirement, which must never be overstated, it often becomes cost-effective to distinguish between a material for electric screening purposes and another material which provides the physical support and determines the mechanical integrity of the screened enclosure.

As an example, consider a plastic box which provides mechanical and, perhaps, environmental protection to an enclosed electronic circuit. This box might be lined with flexible laminates, electroless plating, conductive paints, metallic foil tapes, wire spray or vacuum metallizing. The

MATERIALS FOR SCREENS

Sheet metal
Adhesive metal foil sheet and tape
Flexible laminates
Conductive paint
Wire spray (eg zinc)
Vacuum metallizing
Electroless plating

Figure 12

box might be made of conductive plastic.

Larger screened enclosures are often made of steel-faced wooden sheets or of welded steel sheets mounted on a structural framework.

The final choice will depend on considerations involving the ability to make effective joints to the screening material for items such as access panels, connectors and windows; the avoidance of significant galvanic corrosion; the ability to withstand whatever external environment is stipulated, including mechanical shock and vibration. All these factors must be considered against the cost of achieving the stated required performance.

INTEGRITY OF A SCREENED ENCLOSURE

It has been shown that good screening effectiveness can generally be achieved by reasonably thin metal screens but it is assumed that the screen is continuous and fully surrounds the sensitive item, without gaps or apertures. In practice it is rarely possible to construct a screen in this way. The screen may have to be fabricated in pieces which must be joined together. It may be necessary to penetrate the screen to mount components.

Any decrease in the effective conductivity of the screen, because of joints, will reduce screening effectiveness. Any slots or apertures can act as antennas allowing RF energy to leak in or out. Figure 13 lists some of the reasons why screened enclosures may require joints or apertures.

Now consider, briefly, the attenuation of EM waves through a metallic gap or hole.

REASONS FOR JOINTS OR APERTURES IN SCREENED ENCLOSURES

Seamless construction not feasible
Access panel needed for equipment
　　installation/maintenance
Door for instant access
Ventilation openings needed
Windows needed for viewing displays and meters
Panel mounting components, eg:
　　Connectors for power and signal leads
　　Indicator lamps
　　Pushbuttons　　　Fuses
　　Switches　　　　Control shafts

Figure 13

GAPS AND HOLES IN SCREENS

Concerning the gap or hole which penetrates the screen, as a waveguide through which EM energy is flowing. If the wavelength of this energy is too long compared with the lateral dimensions of the waveguide, little energy will pass through. The waveguide is said to be operating beyond cut-off.

Figure 14 shows formulae for cut-off frequency in round and rectangular waveguide. For operating frequencies much less than the cut-off frequency the formulae for shielding effectiveness are also given. Notice that the attenuation well below cut-off depends only on the ratio of length to diameter. Attenuation of about 100dB can be obtained for a length to diameter ratio of 3. Thus it may be possible to exploit the waveguide properties of small holes in thick screens where penetration is essential. An alternative way of achieving a good length/diameter ratio is to bond a small metallic tube of appropriate dimensions, normal to the screen.

This theory and its extension to multiple holes, forms the design basis for commercially available perforated components such as viewing and ventilation panels which must have good screening effectiveness.

WAVEGUIDE CUT-OFF FREQUENCY (fc)

In round guide, fc	=	175.26/d GHz (6.9/d in.)
d	=	waveguide diameter (mm)
In rectangular guide, fc	=	149.86/a GHz (5.9/a in.)
a	=	largest dimension of waveguide cross-section (mm)

SHIELDING EFFECTIVENESS (S) OF WAVEGUIDE

For operating frequencies well below cut-off

S (round)	=	32.t/d (dB)
S (rectangular)	=	27.2.t/a (dB)
t	=	Screen thickness

Figure 14

SEAMS AND JOINTS

For joints between sheets which are not required to be parted subsequently, welding, brazing or soldering are the prime choices. The metal faces to be joined must be clean to promote complete filling of the joint with conductive metal.

Screws or rivets are less satisfactory in this application because permanent low impedance contact along the joint between the fastenings is difficult to ensure.

For joints which cannot be permanently made, conducting gaskets must be used to take up the irregularities in the mating surfaces. Consideration should be given to the frequency and circumstances in which such joints will be opened and closed during the life of the equipment. One classification defines Class A, B and C joints. Class A is only opened for maintenance and repair. In a Class B joint the relative positions of mating surfaces and gasket are always the same, eg hinged lids and doors. In a Class C joint the relative positions of mating surfaces and gasket may change, eg a symmetrical cover plate.

A wide range of gasket materials is available commercially. They include finger strip; wire mesh with or without elastomer core; expanded metal and oriented wire in elastomer and conductive elastomers. Most suppliers provide estimates of screening effectiveness which can be achieved with the various gaskets. The gaskets come in a variety of shapes to suit many applications. The selection of a suitable gasket depends on many factors, the most important of which are listed in Figure 15.

SOME FACTORS GOVERNING CHOICE OF GASKET

Screening effectiveness
Class A, B or C joint
Mating surface irregularity
Gasket retention method
Flange design
Closure pressure
Hermetic sealing needed?
Corrosion resistance
Vibration resistance
Temperature range
Subject to EMP?
Cost

Figure 15

Appendix 12
Worldwide minimum external noise levels

The figures reproduced below give the *minimum* levels of external noise ever likely to be encountered at a terrestrial receiving site. They are thus a useful guide to the receiver designer, in that there is, in general, no point in designing a receiver to have a noise level much lower than that to be expected from a reasonably efficient aerial system. (The only exception is where, for some special purpose, a very inefficient aerial must be used, e.g. a buried antenna servicing an underground bunker.)

The figures cover the whole frequency range of radio frequencies with which this book is concerned, 10 kHz to 1 GHz, and beyond. The report from which they are reproduced also covers frequencies from 10^{-1} Hz to 10^4 Hz and 1 to 100 GHz.

Figures A12.1 and A12.2 are reproduced from Report 670 (Mod F) 'Worldwide Minimum External Noise Levels, 0.1 Hz to 100 GHz', with prior authorization from the copyright holder, the ITU. Copies of this and other reports and recommendations may be obtained from:

International Telecommunication Union
General Secretariat, Sales and Marketing Service
Place des Nations, CH, 1211 Geneva 20 Switzerland
Telephone: +41 22 730 61 41 (English)/ +41 22 730 61 42 (French)
Telex: 421 000 ut ch/Fax: +41 22 730 51 94
X....400: S=Sales; P=itu; C–ch
Internet: Sales@itu.ch

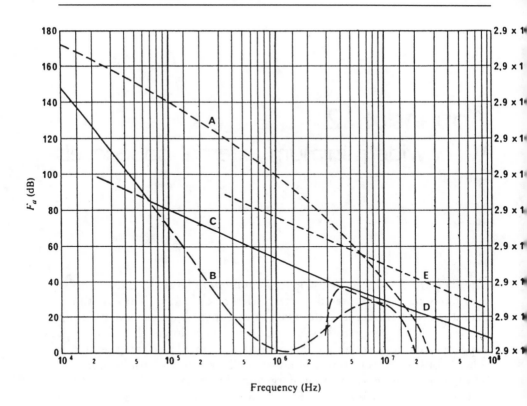

Figure A12.1 F_a *versus frequency (10^4 to 10^8 Hz). This figure covers the frequency range 10^4 to 10^8 Hz, i.e., 10 kHz to 100 MHz. The minimum expected noise is shown via the solid curves and other noises that could be of interest as dashed curves. For atmospheric noise, the minimum values expected are taken to be those values exceeded 99.5% of the time and the maximum values are those exceeded 0.5% of the time. For the atmospheric noise curves, all times of day, seasons, and the entire Earth's surface has been taken into account. More precise details (geographic and time variations) can be obtained from Report 322. The man-made noise (quiet receiving site) is that noise measured at carefully selected, quiet sites, world-wide as given in Report 322. The atmospheric noise below this man-made noise level was, of course, not measured and the levels shown are based on theoretical considerations. Also shown is the median expected business area man-made noise.*

A Atmospheric noise, value exceeded 0.5% of time; B Atmospheric noise, value exceeded 99.5% of time; C Man-made noise, quiet receiving site; D Galactic noise; E Median business area man-made noise, Minimum noise level expected.

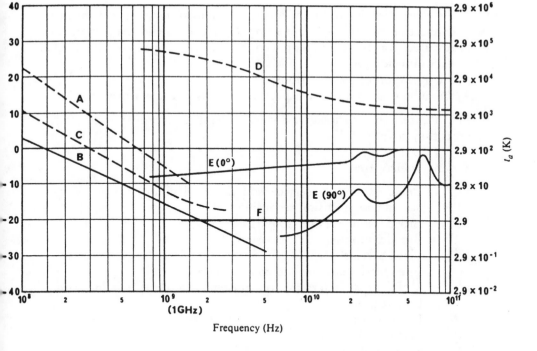

Figure A12.2 F_a *versus frequency (10^8 to 10^{11} Hz). The frequency range 10^8 to 10^{11} Hz is covered, i.e., 100 MHz to 100 GHz. Again, the minimum noise is given by solid curves, while some other noises of interest are given by dashed curves.*

A Estimated median business area man-made noise; B Galactic noise; C Galactic noise (toward galactic centre with infinitely narrow beamwidth); D Quiet sun ($\frac{1}{2}$ degree beamwidth directed at sun); E Sky noise due to oxygen and water vapour (very narrow beam antenna); upper curve, 0° elevation angle; lower curve, 90° elevation angle; F Black body (cosmic background), 2.7 K, Minimum noise level expected.

Annex 1: ITU-R Recommendations and Reports

ITU-R Recommendations constitute a set of standards previously known as CCIR Recommendations. They are the result of studies undertaken by Radiocommunication Study Groups on:

• the use of radio frequency spectrum in terrestrial and space radiocommunication including the use of satellite orbits:

- the characteristics and performance of radio systems, except the inter-connection of radio systems in public networks and the performance required for these interconnections which are part of the ITU-R Recommendations;
- the operation of radio stations;
- the radio communication aspects of distress and safety matters.

ITU-R Recommendations are divided into series according to the subject areas they cover as follows:

Series   Subject area

BO*   Broadcast satellite service (sound and television)
BR   Sound and television recording
BS*   Broadcasting service (sound)
BT*   Broadcasting service (television)
F   Fixed Service
IS   Inter-service sharing and compatibility
M*   Mobile, radiodetermination, amateur and related satellite service
P*   Propagation
RA   Radioastronomy
S   Fixed-satellite service
SA   Space applications
SF   Frequency sharing between the fixed-satellite service and the fixed service
SM   Spectrum management techniques
SNG   Satellite news gathering
TF   Time signals and frequency standard emissions
V   Vocabulary and related subjects

*Also includes ITU-R Reports

There are currently 594 ITU-R Recommendations in force. ITU-R Recommendations are progressively being posted on TIES and will be accessible by subscribers to the ITU-R Recommendations Online Service. For further information please contact the ITU Sales Service.

Appendix 13
Frequency allocations

Frequency allocations are settled on a worldwide basis by WRC, the World Radio Conference, previously known as WARC, the World Administrative Radio Conference. The Conference, which is convened as necessary (usually every two or three years), is held under the aegis of the International Telecommunications Union (ITU), which is itself an organ of the United Nations. Implementation is down to individual countries, not all of which are represented at the WRC, while not all of those that are observe all of the allocations.

Annexe 1: Radio frequency spectrum management in the UK (part of Region 1)

In the UK, frequencies are allocated by the Radiocommunications Agency, which is an Executive Agency of the Department of Trade and Industry. The allocations are contained within the United Kingdom Table of Frequency Allocations, which is in five parts, as follows:

Part 1 RA306   May 1996   9 kHz–28 MHz

Part 2 RA255 (Rev 2) May 1996 ISBN 1855691817 28–470 MHz

Part 3 RA265 (Rev 2) May 1996 ISBN 1855691957 470–3600 MHz

Part 4 RA278 (Rev 1) May 1996 ISBN 185569218X 3600 MHz–30 GHz

Part 5 RA253 (Rev 3) May 1996 ISBN 1855692008 30–105 GHz

These information sheets provide a comprehensive breakdown of the allocated use of the radio spectrum in the UK. However, the information given is an overview of the current use. It is not exhaustive in detail nor does it confer or imply any guarantee that allocations will not be altered in future.

Space limitations prevent the inclusion here of these documents in their entirety, but to give a flavour of the wealth of information contained in them, some sample pages are reproduced below, namely:

Part 1   page 13
Part 2   pages 1 and 12
Part 3   page 3.

The Agency may be contacted as indicated on page 1 of Part 2.

Also of importance to UK readers is Radiocommunications Agency Information Sheet RA168 (Rev 3), May 1996 ISBN 1855690705, entitled 'The European Standards Institute (ETSI) and Radio Standards in the UK'.

Allocation to United Kingdom Services	Comments
6 525–6 685 kHz AERONAUTICAL MOBILE (R)	Civil and non civil aeronautical communication services. British Telecom on 6 634 kHz from Rugby. NATS joint use of 6 622 kHz located in the Republic of Ireland.
6 685–6 765 kHz AERONAUTICAL MOBILE (OR)	Government use.
6 765–7 000 kHz FIXED Land Mobile RR524, 525	Government use.

7 000–7 100 kHz
 	
AMATEUR
AMATEUR-SATELLITE
 	
RR510, 526, 527
	Amateur service operates on a primary basis in the following sub-bands:
	
	   7 000–7 035 kHz CW
	
	   7 035–7 040 kHz datamode, sstv, fax, CW
	
	   7 040–7 045 kHz datamode sstv, fax, phone, CW
	
	   7 045–7 100 kHz phone, CW

7 100–7 300 kHz BROADCASTING	BBC broadcasting services operate within the bands 7 105–7 295 kHz, the channel spacing is 5 kHz and channel bandwidth is 10 kHz. All services are transmitted via Rampisham, Skelton and Woofferton.
7 300–7 350 kHz BROADCASTING RR521A, 521B, 528A	BBC broadcasting services operate on 7 320 kHz and 7 325 kHz in this band via Rampisham, Skelton and Woofferton.
7 350–8 100 kHz FIXED Land Mobile RR529	Government use. NATS data links with Iceland
8 100–8 195 kHz FIXED MARITIME MOBILE	Government use.
8 195–8 815 kHz MARITIME MOBILE RR501, 500A, 500B, 520B, 529A	Government use. Internationally allocated band for the maritime mobile service for two and single frequency single-sideband operation. BT currently operate 38 services within this band. 8 414·5 kHz is used for international distress for digital selective calling. 8 376·5 kHz is used for international distress for narrow-band direct-printing telegraphy. 8 364 kHz can be used for search and rescue.

13

UK USE OF THE BAND 28-470 MHz

Note that the "Allocations to Services" column entries in upper-case denote PRIMARY status and those in lower-case denote Secondary status as defined in the ITU Radio Regulations. Footnotes have been included (with a RR prefix) where they apply to the UK-details of the relevant footnotes are given in Annex B.

INDEX

Main Table   Page 2

Annex A-List of Abbreviations.   Page 19

Annex B-List of Relevant ITU Footnotes.   Page 21

Annex C -CEPT Recommendations and UK Specifications.   Page 25

Annex D -List of Relevant Specifications   Page 28

For further information on radio matters contact the
RA 24 hour Enquiry Point Service 0171-211 0211 or write to
Radiocommunications Agency
New King's Beam House, 22 Upper Ground, London SE1 9SA

1

© Crown Copyright Radiocommunications Agency.

Allocation to United Kingdom Services	Comments
169·39375-169·81875 MHz LAND MOBILE RR613, RR613B	Paging, ERMES, within the range 169·4125-169·8125 MHz, specifications ETS 300/133/1-7, ETS 300/340, TBR 007 and ETR 050 apply.
169·81875-169·84375 MHz LAND MOBILE RR613, RR613B	PMR single frequency systems operate within the range 169·81875-169·84375 MHz, specifications MPT 1302, MPT 1303, MPT 1304 and MPT 1326 apply. Specifications ETS 300 086, I-ETS 300 113 and I-ETS 300 219 also apply.
169·84375-173·04375 MHz LAND MOBILE RR613, RR613B	PMR operates within the range 169·84375-173·04375 MHz, specifications MPT 1302, MPT 1303, MPT 1304 and MPT 1326 apply. Specifications ETS 300 086, I-ETS 300 113 and I-ETS 300 219 also apply.
173·04375-173·09375 MHz LAND MOBILE RR613, RR613B	PMR single frequency systems operate within the range 173·04375-173·09375 MHz, specifications MPT 1302, MPT 1303, MPT 1304 and MPT 1326 apply. Specifications ETS 300 086, I-ETS 300 113 and I-ETS 300 219 also apply.
173·09375-174·0 MHz LAND MOBILE RR613, RR613B	Home Office/Scottish Office for the Emergency Services. SRDs operate on 173·1875 MHz, specifications MPT 1360 applies. They also utilise 173·1875 MHz, specification MPT 1344 applies, see also RA 114. These services are exempt from licensing. SRDs operate in the bands 173·20-173·35 MHz and 173·7-174·0 MHz, specifications MPT 1328, MPT 1330 and MPT 1312 apply, see also RA 114. These services are not exempt from licensing. Radio aids for the Deaf operate in the range 173·35-174·0 MHz and wide band radio microphones operate in the range 173·8-174·0 MHz, specification MPT 1345 applies to both applications and both are exempt from licensing. See also RA 114. Indoor radio microphones operate within the bands 173·70-174·0 MHz, specification MPT 1350 applies.

12

Table of Current Allocations and Users
in the Range 470-3600 MHz

Allocation to United Kingdom Services	Comments

470-590 MHz
 	
    BROADCASTING
 	
    Land Mobile
 	
    RR677A
	TV Broadcasting Band IV; TV Channels 21-35. Some SAB use.
	
	CEPT Recommendation T/R 72-01 refers.

590-598 MHz
 	
    AERONAUTICAL
      RADIONAVIGATION
 	
    Land Mobile
 	
    RR677A, 686, 686A.
	Airport radars now confined to TV Channel 36 where they are permitted in the UK by footnote RR686.
	
	CEPT Recommendation T/R 72-01 refers.

598-854 MHz
 	
    BROADCASTING
 	
    FIXED
 	
    Land Mobile
 	
    RR677A, 689, 695A
	TV Broadcasting Band V; TV Channels 37-68. Some SAB use.
	
	Radio astronomy use in the sub-band 606-614 MHz (TV Channel 38); footnote RR689 refers. Used for pulsars and VLBI.
	
	No UK FIXED use.
	
	CEPT Recommendation T/R 72-01 refers.

854-862 MHz
 	
    FIXED
 	
    Land mobile
 	
    RR695A.
	TV Channel 69.
	
	Used mainly in the UK for SAB (mostly radiomicrophones operating to specification MPT1350).
	
	Some military use, mainly transportable radio relays.
	
	CEPT Recommendation T/R 72-01 refers.

862-864 MHz
 	
    FIXED
 	
    MOBILE except aeronautical
      mobile
	Some use by the Home Office and Scottish Office for the Emergency Service.
	
	862·25-864 MHz set aside for CT2 expansion.
	
	CEPT Recommendations T/R 25-08, T/R 72-01, T/R 75-02 refer.

3

Annexe 2: Radio frequency spectrum management in the US (part of Region 2)

The Communications Act of 1934 provides the foundations for US spectrum rules and regulations, management and usage. The basic authority is delineated in Sections 303, 304 and 305 of the Act. Section 303 presents the general powers of the Federal Communications Commission (FCC) regarding transmitting stations; 304 deals with waiving frequency claims; and 305 provides that Federal Government owned stations shall be assigned frequencies by the President (delegated to the Department of Commerce National Telecommunications and Information Administration [NTIA] via Executive Order 12046). Section 305 is particularly significant as it provides for the separation of authority between the Federal Government and the non-Federal Government, or private sector. Section 305 has resulted in two US spectrum regulatory bodies: the FCC regulating the non-Federal Government sector, and the NTIA regulating the Federal Government sector. Section 305 has also resulted in agreements between the Federal Government and non-Government sectors that essentially divide the spectrum usage into three parts: exclusive Federal Government use, exclusive non-Federal Government use, and use shared between the two sectors.

The NTIA is aided by other federal agencies and departments through an advisory group, the Interdepartmental Radio Advisory Committee (IRAC). IRAC carries out frequency coordination for the Federal Government Agencies, recommends technical standards, and reviews major Federal Government systems to assure spectrum availability. The IRAC also provides advice to the NTIA on spectrum policy issues.

Although the NTIA and FCC generally operate independently of each other, they coordinate closely on spectrum matters. An FCC liaison representative participates in the IRAC, and the NTIA participates in the rule making process of the FCC with the advice of the IRAC. FCC and NTIA spectrum sharing coordination is also carried out daily as required.

For the purposes of international coordination, the ITU divides the world into three regions as presented in Figure A13.1, with each region having its own allocations, although there is much commonality among the regions. Each region has over 400 distinct frequency bands and hundreds of footnotes (exceptions or additions to the table). Also reproduced (as Table A13.1, below) is a sample page from the frequency allocation table as it applies internationally, and to the US in particular.

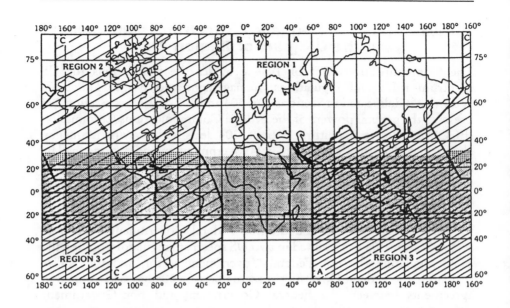

Table A13.1 *Sample of frequency allocation table.*

Table A13.1 Regions defined for frequency allocations. Shaded area represents tropical zone.

| INTERNATIONAL | | | UNITED STATES | | | | |
Region 1 MHz	Region 2 MHz	Region 3 MHz	Band MHz 1	National Provisions 2	Government Allocation 3	Non-Government Allocation 4	Remarks 5
3600–4200 FIXED FIXED-SATELLITE (Space-to-Earth) Mobile 781 782 785	3500–3700 FIXED FIXED-SATELLITE (Space-to-Earth) MOBILE except aeronautical mobile Radiolocation 784		3500–3600	US110	AERONAUTICAL RADIONAVIGATION (Ground-based) RADIOLOCATION G59 G110	Radiolocation	
			3600–3700	US110 US245	AERONAUTICAL RADIONAVIGATION (Ground-based) RADIOLOCATION G59 G110	Radiolocation FIXED-SATELLITE (Space-to-Earth)	
	786						
	3700–4200 FIXED FIXED-SATELLITE (Space-to-Earth) MOBILE except aeronautical mobile 787		3700–4200			FIXED FIXED-SATELLITE (Space-to-Earth) NG41	
AERONAUTICAL RADIONAVIGATION 789 788 790 791			4200–4400	US261 791	AERONAUTICAL RADIONAVIGATION	AERONAUTICAL RADIONAVIGATION	
4400–4500 FIXED MOBILE			4400–4500		FIXED MOBILE		
4500–4800 FIXED FIXED-SATELLITE (Space-to-Earth) MOBILE 792A			4500–4800	US245	FIXED MOBILE	FIXED-SATELLITE (Space-to-Earth)	

Appendix 14
SRDs (Short Range Devices)

The area of wireless SRDs is currently one of explosive growth, with new applications appearing almost daily. Typical examples are wireless LANs, radio microphones, telemetry, garage door openers, radar level gauges, radio-operated car security keys, etc. In many cases (but not currently in the UK radar level gauges and the higher powered versions of radio microphones and tagging devices), the key feature is that no licence is required by the end-user. But of course the manufacturer or importer is responsible for ensuring that the device meets the appropriate national mandatory regulations. The regulations vary from country to country, and approval is usually needed in any country in which a supplier proposes to sell his devices.

In Europe, the regulations are in the process of being harmonized throughout the EU, under the aegis of ETSI. The European Telecommunications Standards Institute aims to coordinate its activities with other relevant standards and international bodies such as CEPT, CEN/CENELEC, EBU, ITU/CCITT/CCIR and ITSTC. ETSI produces ETSs (European Telecommunications Standards) and I-ETSs (Interim ETSs), as well as ETRs (ETSI Technical Reports). A useful introduction to ETSI is contained within the information sheet 'The European Telecommunications Standards Institute (ETSI) and Radio Standards in the UK', which has already been mentioned in Appendix 13.

Information on the standards applicable to SRDs in particular (in the UK) is contained in the information sheet 'Short Range Devices', RA 114 (Rev. 5), March 1996, ISBN 1855692058, also available from the Radiocommunications Agency. Pages 6 and 7 of this document are reproduced below. These give details of the relevant UK standards (usually prefixed MPT-), but note that they should be read in conjunction with the Notes on page 8 of the document. These standards will continue to apply in the UK until a relevant ETS is published, which will then supersede the MPT specification.

Manufacturers wishing to supply the European market must ensure that their product meets the national standard in force in any given country, or the relevant I-ETS, whichever is applicable. Most SRDs will be covered by I-ETS 300 220: 1993, which applies to devices operating in the range 25–1000 MHz. With the advent of the European Directive 89/336/EEC, EMC has recently become an additional requirement for SRDs. In most cases, the applicable standard is prETS 300 683: 1995.

TYPES OF SHORT RANGE DEVICES EXEMPT FROM LICENSING			ANNEX 1
Uses	Frequency	Maximum ERP	Specification
Medical and Biological Telemetry			
Medical/Biological Telemetry	300 kHz - 30 MHz	See specification	W6802
Medical and Biological Telemetry (narrow band and wide band)	173.7 - 174 MHz	10 milli Watts	MPT 1312
Medical/Biological Telemetry	458.9625 - 459.1000 MHz	500 milli Watts	MPT 1329*
General Telemetry and Telecommand Devices			
General Telemetry and Telecommand	26.995 MHz 27.045 MHz 27.095 MHz 27.145 MHz 27.195 MHz	1 milli Watt	MPT 1346
Telemetry Systems for Databuoys	35 MHz	250 milli Watts	MPT 1264
General Telemetry and Telecommand (narrow band)	173.2 - 173.35 MHz	10 milli Watts	MPT 1328
General Telemetry and Telecommand (wide band)	173.2 - 173.35 MHz	10 milli Watts	MPT 1330
General Telemetry, Telecommand and Alarms	417.90 - 418.1 MHz	250 micro Watts	MPT 1340
Vehicle Radio Keys	433.72 - 434.12 MHz	10 milli Watts	MPT 1340
Industrial/Commercial Telemetry and Telecommand	458.5 - 458.95 MHz	500 milli Watts	MPT 1329**
Alarms			
Short Range Alarms for the elderly and infirm	27.450 MHz 34.925 MHz 34.950 MHz 34.975 MHz	500 micro Watts	MPT 1338
General Alarms	417.90 - 418.10 MHz	250 micro Watts	MPT 1340
Vehicle Paging Alarms	47.4 MHz	100 milli Watts	MPT 1374
Marine Alarms for Ships	161.275 MHz	10 milli Watts	MPT 1265
Mobile Alarms	173.1875 MHz	10 milli Watts	MPT 1360
Short Range Fixed In Building Alarms between 1 mW and 10 mW	173.225 MHz	10 milli Watts	MPT 1344
Fixed Alarms	458.8250 MHz	100 milli Watts	MPT 1361
Transportable and Mobile Alarms	458.8375 MHz	100 milli Watts	MPT 1361

6

Use	Frequency	Maximum ERP	Specification
Alarms (cont)			
Vehicle Paging Alarms with integral Radio Key	458.9000 MHz	100 milli Watts (paging) 1 milli Watt (radio key)	MPT 1361
Model Control			
General Models	26.96 - 27.28 MHz	100 milli Watts	N/A +
Air Models	34.955 - 35.255 MHz	100 milli Watts	N/A +
Surface Models	40.665 - 40.955 MHz	100 milli Watts	N/A +
General Models	458.5- 459.5 MHz	100 milli Watts	N/A +
Short Range Microwave Devices or Doppler Apparatus		**Maximum EIRP**	
Apparatus designed solely for outdoor use	10.577 - 10.597 GHz	1.0 Watt	MPT 1349
Apparatus designed for indoor use and Short range data links within one building	10.675 - 10.699 GHz	1.0 Watt	MPT 1349
Apparatus Designed for fixed or portable applications	24.150 - 24.250 GHz	2.0 Watts	MPT 1349
Apparatus designed solely for use in a mobile application	24.250 - 24.350·GHz	2.0 Watts	MPT 1349
Anti-Collision Devices	31.80 - 33.40 GHz	5.0 Watts	MPT 1349
Any apparatus not within any category above and short range data links within one building	2.445 - 2.455 GHz	100 milli Watts	MPT 1349
Other Devices			
Spread Spectrum Applications (including Radio Lans)	2.4 - 2.483 GHz	100 milli Watts	ETS 300 328
Induction Communication Systems	0 -185 kHz and 240 - 315 kHz	See specification	MPT 1337
Metal Detectors	0 -148.5 kHz	See SI 1980No1848	N/A +
Access and Anti-Theft Devices and Passive Transponder Systems	2 - 32 MHz	See specification	MPT 1339
Teleapproach Anti-Theft Devices	888-889 MHz	See specification	MPT 1353
Teleapproach Anti-Theft Devices	0 -180 kHz	See specification	MPT 1337
General Purpose Low Power Devices	49.82 - 49.98 MHz	10 milli Watts	MPT 1336
Cordless Audio Equipment	36.61 - 36.79 and 37.01 - 37.19 MHz	10 micro Watts	MPT 1336
Radio Microphones	174.600 - 175.020 MHz	5 milli Watts (narrowband)	MPT 1345
Radio Microphones	173.800 - 175.000 MHz	2 milli Watts (wide band)	MPT 1345
Radio Hearing Aids	173.350 - 174.415 MHz	2 milli Watts	MPT 1345

Index

Absorption 184, 185
Accumulator 123
Admittance 11, 237–240
ADC 170
AFC 82, 100, 167
AGC (automatic gain
    control) 64, 79, 82, 97,
    166, 171
Air gap 35
Ambient 142
Ampere 4
Amplifier 10, 204, 215
    buffer 107 , 216, 217
    cascode 64, 77
    class A 130
    class B 130
    class C 11, 95, 130
    class D 110
    IF 82
    limiting 83
    log 83
    low noise 72
    parametric 188
    push pull 67, 131
    RF 73–84, 188
        power 129–159
    single ended 131
    small signal 73–84
    video 215
AMPS 228
Angle 180
    conduction ~ 138
    take-off ~ 198
Angular velocity 5
Antenna 160, 190–211

active 204–207
Adcock 210
aperture, effective 199,
    204, 209
array
balanced 27
dipole 177, 179, 190,
    195, 198, 199, 204
    0bandwidth of 193
    impedance of 193
directional 187, 188
discone 193
dish 209, 210
doublet 190
gain 191
ideal isotropic 177, 179,
    181, 190, 204, 232
log periodic 209, 232
loop 199, 207
measuring 231
monopole 196, 204
omnidirectional 209
resonant, non resonant
    190
rhombic 29
sidelobes 193, 196, 197
tuning unit (ATU) 190,
    193
Yagi 194, 209
Area
    capture, cross section,
        intercept see Antenna,
        aperture
Argument see Phase
ASCII (ITA5) 91, 94

ASIC 227
Asynchronous code 91
Atmosphere 183
Atomic lattice 4
Attenuation 15, 16, 21, 23, 44
Attenuator 212–215
    balance pad 34–6
    pad
        fixed 215
        mismatch 215
        minloss 210
    PIN diode 131
    piston 213
    variable 213, 215
ATU see Antenna
Audio 98, 100
Aurora Australis, Borealis
    187
Automatic frequency control
    see AFC
Automatic gain control see
    AGC
Automatic repeat request
    (ARQ) 92

Balance 21, 32 35
    floating 34
    measurements 34
    ratio 35
Balun 33, 38–42, 193
Bandpass 15, 75
    filter 72
    roofing 72
Bandwidth 80, 85, 165, 178,
    193, 199

instantaneous 193, 209
noise 216
occupied (OBW) 91, 95,
    216
post detector (video) 223
3dB 15, 178
Baseband 85, 88, 97
composite, signal 160
Battery 4
Baud 91, 94, 96
Beryllium oxide 129
Bias *see* Circuit
Blocking 165
Bode plot 115, 226
Boltzmann's constant 80
Bright-up 229
Butterworth 16, 132

Cable, coaxial *see* Feeder
CAD (computer aided
    design) 73
Capacitance 1, 4–7, 11, 54
feedback (reverse transfer)
    64
inter-electrode 79
interwinding 29
self 29, 30
stray 5, 155
winding 40
Capacitor 4–7, 14
ceramic 5, 135
    NP0 7, 135
decoupling 6, 7
electrolytic 6
ganged 15
mica 135
polystyrene 5
porcelain 135
trimmer 16
variable 7, 54
Capture effect 171
Carrier 55, 97
lifetime 55
recovery circuit 229
majority 61
minority 56
sub-, suppressed 160
wave 85
Cartesian 137
CCIR *see* ITU
Channel 116
spacing 164
Charge 4–7
Chebychev 16, 38, 132
Chipping sequence 101

Chirp
linearity 223
sounder 185
Choke (RF) 10, 138
sectionalized 11
Circle diagram 12
Circuit 4, 10
active 143
bias 143
common (grounded) base
    61, 73, 78
collector 61, 77
drain
emitter 59, 75, 78
gate 75
open 30, 47, 59, 157
primary 15, 27
resonant *see* Resonance
secondary 15, 27
short 30, 39, 47, 59, 157
source
tuned 14
CISPR 231
Clarifier 85
CMOS
Coefficient
coupling 15
    capacitive 15
    inductive 15
reflection 23, 26, 225, 226
temperature *see* Tempco
Coercivity 32
Combiners *see* Couplers
Common mode *see* Mode
Conductance 1, 2, 21,
    237–240
mutual 73
output 75
Conductivity 182, 198
Conductor I
Constantan 2
Convolver
SAW (surface acoustic
    wave) 101
Core
E 35
pot 35
Corkscrew rule 7
Coulomb 4
Couplers 44–51
directional 49–51, 131
hybrid 45
    sum and difference 46
quadrature 50
resistive 44

Coupling
critical 15
thermal 144
Crystal, quartz 18
AT cut, OCXO, TCXO
    18, 110
BT cut 112
doubly rotated 112
filter 19, 38
fundamental frequency of
    109
monolithic dual 19
pulling 19
SCcut 19, 112
singly rotated 112
strain compensated *see* SC
    cut
Current 1
bias 138
constant generator 14, 15
conventional 4
consumption 138
gain (transistor) 58
forward-transfer ratio *see*
    Current gain
magnetizing 27, 40
CW (continuous wave) *see*
    Signal

DAC 123
Damping 108
frequency selective 154
Data
clock 187
rate 187
transmission 185
dc 5
conditions 58
DBM *see* Mixer
Dead
space 121
zone 183
Decoupling 10
Demodulation 97, 221
synchronous 229
Depletion layer 54, 55, 61
Detector
diode 50, 97, 131
Law assessor 92
phase 115
    EXOR 121
    logic 121
    pump up/pump down
        121
    sample and hold 121

quadrature 82, 99
ratio 98, 100
slide-back 92
Deviation 88
DF (direction finder/ing)
    186, 210, 211
Dielectric 5
    constant (see also
        Permittivity) 182
Differential
    encoding 94
    decoding 95
Diffraction 181
Digital to analog converter
    see DAC
Diode 52, 144, 220
    avalanche 55
    noise 56
    PIN55
    schottky (hot carrier) 55,
        70
    snap-off 55
    variable capacitance see
        Varactor
    zener 55
Dipole see Antenna
Direct
    current see dc
    sequence 101, 187
Direction finder/finding, see
    DF
Directivity 191, 204, 210
Dissipation 142
Distortion 32, 54, 65, 118
    phase
    second order 66, 68
    third order 66, 68
Divider
    variable ratio 79, 114
Doppler 121
Doublet see Antenna
Doughnut
Drift 104
DSP (digital signal
    processing)
DTMF 228
Ducting 182, 183
Dust iron 11
DX 40, 85
Dynamic range 80

Eavesdropping 1
Efficiency 138
    antenna 190
Egli 181, 188

Electrodes 18
Electromagnetic
    compatibility see EMC
Electromotive force see EMF
Electron 4, 52
Elliptic 18, 132
EMF I, 4, 5, 7, 9, 199
    back 9, 27
EMC 229
Emitter follower see
    Common Collector
    Circuit
End effect 191
Equaliser 229
Equivalents
    delta/star 235
    series/parallel 234
ESR 135
Eye diagram 221, 229
    closure 221

Fading 166
    selective 166
FCC 225, 231
Feedback 147
    internal 61, 75
    lossless 82
    negative 82
Feeder 194, 204
    balanced 194
    coaxial 21, 40
    flat twin 21
Ferrite 7, 11, 32, 35
    saturation of 35
FET see Transistor
Field
    electric (E) 177, 208, 233
    far 178, 179, 233
    induction178, 179
    lines; electric, magnetic
        177
    magnetic (M) 7–11, 177,
        208, 233
    near 177, 179, 233
    probes (E & H) 231
    radiation 178
    strength 219, 229
FIFO 101
Filter, types of see type; e.g.
    Butterworth, Elliptic,
    etc.
all-pass (see also
    equalizer) 95, 215
bandpass 15, 19, 131
Bessel 216

Caur see Elliptic
FIR (finite impulse
    response) 216
highpass 12
linear phase 216
lowpass 11, 131, 168
harmonic 130
ripple 18
roofing 217
Flicker 105
Flipflop 168
Flux
    density 35
    magnetic 7–11, 27, 40
FM see Modulation
FOT (frequence optimum de
    transmission) 184
Frequency 5, 15
    antiresonant 19
    clock 165
    comparison
    sidebands 79
    cut-off 132
    division 154
    hopping 100
    lock loop (FLL) 91, 114
    meter, digital
    multiplication 154
    multiplier 54
    resonant 14, 15, 19
    unity (loop) gain 118

Gain 108, 142
    directional
    compression 68
    current 58
    open loop 120
    processing 101,187
    stage 68, 73, 76
Gaussian distribution 103
Generator 14, 21
    constant
        current 14, 248
        voltage 248
    ISM 187
    signal 21, 114
Germanium 52
Ghosting 210, 219
Goniometer 210
GPIB 221
Great circle 186
Ground plane 155, 196
Group delay 216

Harmonics 10, 19, 55, 109,

155, 164, 219, 221
Heat sink 142, 155
HF 75, 91, 101, 135, 182,
    184–186, 199, 201,
    2205, 207, 209, 210,
    220, 223, 231
High frequency see HF
High pass 12, 38
Hole 53
Hybrid
    coupler see Coupler
    thick film 130
Hyperbolic equations

IC 72, 82
IF 82, 89, 97, 101, 162
Image response see Response
Impedance 11, 237–240
    characteristic 21, 213, 225
        of free space 178, 233
    complex conjugate 137
    imaginary 137
    input 73
    real 137
Inductance 1, 7–11
    leakage 29, 38, 40
    magnetizing 27
    mutual 15
    primary 27, 30
    self 2
    stray series 155
    tempco of 37
Inductor 14, 35
Insulator 1, 4
Integrated Circuit see IC
Integrator 118
Intercept point 68
Interference 171, 173, 188,
    210, 274
Interlocks 160
Intermediate frequency see IF
Intermodulation 32, 44, 165,
    223
    back see Reverse
    reverse 158
    third order products 68
Intersymbol interference see
    ISI
Inverse square law 179
Ionosphere 183
    ionospheric scatter 186
    storms 185
ISI 92, 95
ISM (industrial, scientific and
    medical) see Generator

Isolation 44
    reverse 78
Isotropic, see radiator
ITA2 91, 94
ITA5 see ASCII
ITU 188, 189, 225

j 6, 9
Jammer, jamming 91, 188
Jitter 95
Joule 1, 5

Layout 68
Lenz's Law 9
Limiter
    hard 100
    low phase shift 83
Linearity 2, 65, 80
    non- 129, 154, 164
Line stretcher 157
LO (Local oscillator) 64, 70,
    72
Lock box 114
Logarithmic 12
Loss
    absorption 184, 185
    conversion (of mixer)
        70
    core 32
    eddy current 27, 32, 136
    hysteresis 27, 32
    insertion 30, 35
    internal 44
    path 179, 181, 182
    propagation 179
        in free space 178, 233
    radiation 136
    reflection
    resistive 70, 136
    return 215
    series 54
Low pass 12, 16
LUF (lowest usable
    frequency) 184, 185

Magnetomotive force see
    MMF
Magnitude 12, 13
Manganin 2
Maser 80
Matching 22, 23, 26, 27
    delta match 193
    input 150
    matched load 22
    output 137

Maximum available gain
    (MAG) 77
Maximum unilateralized gain
    (MUG) 77
Maxwell's equations 190
Meteorscatter 186
Microcontroller 165
Microstrip 50
Mismatching 77
Mixers 70, 167
    double balanced (DBM)
        70, 72, 164, 216
    image reject 168
    ring 70
MMF 7
Mode
    ambush 211
    common 194
    normal 194
    push-pull, push-push 194
    transverse 194
    zero sweep 223
Modems 96
Modulation 85–102
    AM 54, 85, 97, 130, 164
    APK (QAM) 97
    classification of 257, 258
    CPFSK 91
    cross 165
    CW 40, 85, 167, 185
    FEK 91
    FSK 91, 167
    FM 73, 88, 98, 100, 130,
        160
    incidental 227
    index 88, 117, 229
    ISB 85
    LSB 85
    MFSK 92
    domain 221
    meter 221
    MSK 93
    NBFM 170
    OOK 92
    PM 167
    PSK (BPSK, QPSK, etc)
        92, 101, 221
    QAM (APK) 221
    Serasoidal modulator 160
    SSB 85, 130, 166
    TFM 93
    USB 85
    z ~ 229
Module 130, 162
Modulus see Magnitude

Morse code 85
MUF (maximum usable frequency) 184, 185

Negative feedback see NFB
Nepers 22
Network
    analyser 225–227
    constant resistance 216
    crossover 216
Neutralization 77
NFB 116
NICAM 96
Nichrome 1
NMT 228
Noise 127, 187
    atmospheric 166, 187, 188
    equivalent input 82
    external 197, 188
    flat 105
    figure 79, 166
    floor 105, 118
    galactic 187
    limited receiver 166, 279
    man-made 166, 187
    phase 203, 204, 221, 223, 227
    thermal agitation 80, 188
    triangular spectrum 89
    white 80
Non-linear 52, 54, 77
Normalized values 17, 24
Northern Lights
Notch 40
Nyquist diagram 115

Ohm's law I
OBW see bandwidth
Okamura
Oscillation 76
Oscillator 103–28
    carrier reinsertion 85
    Clapp 106
    clock 110
    Colpitts 77, 105, 115
    crystal 104, 107
    OCXO (oven controlled) 110
    TCXO (temperature controlled) 110
    electron coupled 106
    emitter coupled 107
    filter/amplifier 105
    Franklin 107

Hartley 105
negative resistance 105
Pierce 106
TATG 106
Vakar 111
voltage controlled (VCO) 114, 227
Oscilloscope 228
Outphasing 170
Overtones 19–112
OWF (Optimum working frequency) 184

PA (power amplifier) see Amplifier
Pad see Attenuator
PAL 96
Parallel 2, 12, 14
Parameters
    hybrid 59, 62
    real, complex 59
    s-parameter 59, 137, 205, 241–245
    test set 30
Parity 92
Partial response 97
Patterning 182
Peaking 16
Pentode 58, 64
PEP see Power
Permeability 7, 27, 30
    relative 7
Permittivity 5, 198
    relative 5
Phase 5, 6, 9, 12, 13, 22, 33, 46, 88, 107
    antiphase, inphase 168
    equalizer 95, 215
    lead, lag 6, 168
    lock loop see PLL
    shifter 170
quadrature 94, 99, 168
Piezo-electric 1, 18
PIN see Attenuator
PLL 83, 114, 223,227
PMR (private mobile radio) 116, 126, 181
POCSAG 228
Polarization 180, 190
Potential
    barrier 54
    difference 9
Power 1, 23, 138
    forward, reverse 23, 50, 131

maximum – theorem 22, 199, 235–237
peak envelope 167, 225
    meter 155
    radiated 190
Poynting vector 209
PRBS (pseudorandom bit sequence) 101
Prediction programs (propagation) 184
Pre-emphasis 90
Prescaler
    fixed 116
    two-modulus 79, 115
Primary circuit see Circuit
Privacy device 225
Propagation 177
    constant 22
    error 97
    line of sight (LOS) 181
    multipath 185, 210
Pseudo-Brewster angle 196
PTFE 32

Q (quality factor) 11, 15, 19, 33, 54, 76, 105, 135, 178, 199
Quadrature see Phase
Quarz see Crystal

Radar 179
Radian 5, 117
Radiation 177
    direction of 177
    intensity of 177
    low angle 196
    pattern 190, 194, 206
Radiator
    isotropic 177
Radio
    data see RD
    frequency see RF
    telex 185
Random walk 105
Range laws 179, 181
Ray 180, 181
RD 160
Reactance 6, 9, 11, 13, 14, 237–240
Read only memory see ROM
Receiver 160–76
    communications 75
    homodyne 167
    measuring 229

panoramic
superhetrodyne 164
super-regenerative 167,
170
self-quenched 173
synchrodyne 106, 167
Reflection 180, 181
Reluctance 8, 9, 27
Resistance 1, 2, 8, 11, 13,
237–240
characteristic – of free
space 178
contact 142
dynamic 105
effective 10, 15
equivalent series *see* ESR
incremental (slope) 54, 56,
58
loss 28
negative 105
radiation 178, 190, 199
series 15
thermal 142
winding 29
Resistivity 1, 32
Resistor 1, 9, 14
composition 2
film 2
carbon 2
tin-oxide 2
terminating 47
variable 2
wirewound I
Resonance 14, 107
parallel 107
series 107
Resonator
ceramic 110
Response
IF 165
image 164
spurious 164
RF 21, 27, 52, 54, 73, 85, 97,
129–159, 229
RFI *see* Interference
ROM 123

Safety 129
Saturation 108
saturated output power 68
Scrambler *see* Privacy device
Screen(ing) 34, 68
can 136
SECAM 96
Secondary circuit *see* Circuit

Selectivity 15
Semiconductor 52, 72
doping level 52, 56
intrinsic 52, 54
junction 52
N type, P type 52
point contact 52
stored charge in 55
Series 2, 12, 14
Servo
bang-bang 121
Shunt *see* Parallel
SIDs 186
Sideband 85–102
noise 103, 118
separation 170
unwanted 170
upper (USB) 225
Sidelobes 95, 210
Signal
broadband ~ 193
co-channel 211
CW 85, 185, 187, 219, 228
~ to noise ratio 166, 187,
221
Silicon 52
Sinewave 5, 9, 123
Single sideband *see* SSB
Sinusoidal *see* Sinewave
SITs 186, 188
Skin effect 10, 23
Skip distance 183
Smith chart 24–26, 226,
237–240
SNR *see* Signal
Solar
flares 185
heating 142
Solenoid 7, 11
Solid state 47
Space
free 179, 193
Spectrum 85
analyser 155, 223–225
spectral lines 90
spread 100, 187
Splitters *see* Couplers
Spurious
internal 165
response (spur) 70, 165
spectral line (spur) 123,
165
Spurs *see* Spurious
Squelch 83
Squegg 171, 229

SSB 85, 130, 229
Stability 61, 76, 154, 157
margin 77
Star-delta transform 3
Stereo 89
Stripline 50
Stub 26
Substrate 64
Sun 188, 281
spot cycle 184, 187
Superconductivity 204
Susceptance 6, 9, 237–240
Symbol 92
Sync(hronization) 101
Synthesizer 79, 110
direct 121, 227
digital 121, 227

TACS 228
Tank circuit 68
Tempco 7, 55, 105
Temperature
coefficient *see* Tempco
inversion 182
TEMPEST 274
Test
equipment 219–233
two tone ~ 209
Tetrode 64, 79
Thermal
agitation 80
runaway 58
Thermistor 143
Thermocouple 220
TIDs 186, 189
Time
constant 90, 95, 97
delay 185
propagation 92
settling 223
Tone separation 168
Top cut *see* Low pass
Toroid 7, 9
Transceiver 160–76
Transformer 11, 27–43
inverting 40
line 40
non-inverting 42
quarter wave 23
RF
broadband 137
transmission line 40,
153
Transistor 56–66
bipolar 56, 143

depletion layer 56
field effect
depletion mode 63
drain, gate, source 61
    junction 61
    metal-oxide-silicon
        (MOSFET) 72, 131
depletion, enhancement
    64
dual gate 64, 79, 106
substrate (bulk) 64
RF power 129–59
unipolar 61
Transmission
    characteristic 225
    lines 21–6, 40
Transmitter 29, 47, 160–76
    broadcast 160
    spark 103
TV 164, 182, 210
Triode 64
Trombone 157
Troposcatter 186
Tuned circuit 11, 14, 103
    bandpass 15
    coupled 15, 75
        series 14, 19
        parallel 14, 19
    single 15
    stagger 15
    synchronously 15
TV 40, 42
UHF 54, 101, 107, 116, 135,
    181, 194, 199, 200

Ultra High Frequency see
    UHF
Unbalanced 21
    feeder 27
Unilateralization 77
UV (ultra violet) 185

Vacuum 4
Vapour pressure 104
Varactor diode see Varicap
Varicap (Varactor diode) 54,
    100, 114, 118, 154
    hyperabrupt 54
VCO see Oscillator
VDE 231
Vector 14, 24
Very High Frequency see
    VHF
Very Low Frequency see
    VLF
VHF 4, 73, 88, 101, 116,
    136, 160, 173, 181,
    182, 194, 199, 200,
    209, 210, 231
Video 83
VLF 183
Voltage
    low 138
    over 138
    pinch-off 61
    reverse breakdown 147
    threshold 143
        controlled oscillator
            (VCO) 79

standing wave ratio see
    VSWR
VSWR 23–6, 131–57, 193,
    196, 208, 217

Walkie-talkie 162
Watt 1, 5
Wave
    ground 182, 183
    guide183
    incident, reflected,
        velocity 23
    long, medium, short 164
    plane 178, 199
    spherical 178, 199
        front 178, 182
        impedance 276
        length 193–196
Waveform 5
Wavelength 22
Winding 31
    sectionalization 32
Wire 33
    enamelled, magnet 42
Wullenweber 210
Wye-mesh transform see
    Star-delta transform

X rays 185

Z modulation